Growth and Ecosystem Services of Urban Trees

Growth and Ecosystem Services of Urban Trees

Special Issue Editor
Thomas Rötzer

MDPI • Basel • Beijing • Wuhan • Barcelona • Belgrade

Special Issue Editor
Thomas Rötzer
Technical University of Munich
Germany

Editorial Office
MDPI
St. Alban-Anlage 66
4052 Basel, Switzerland

This is a reprint of articles from the Special Issue published online in the open access journal *Forests* (ISSN 1999-4907) from 2018 to 2019 (available at: https://www.mdpi.com/journal/forests/special_issues/Urban_Trees).

For citation purposes, cite each article independently as indicated on the article page online and as indicated below:

LastName, A.A.; LastName, B.B.; LastName, C.C. Article Title. *Journal Name* **Year**, *Article Number*, Page Range.

ISBN 978-3-03921-592-8 (Pbk)
ISBN 978-3-03921-593-5 (PDF)

Cover image courtesy of Thomas Rötzer.

© 2019 by the authors. Articles in this book are Open Access and distributed under the Creative Commons Attribution (CC BY) license, which allows users to download, copy and build upon published articles, as long as the author and publisher are properly credited, which ensures maximum dissemination and a wider impact of our publications.
The book as a whole is distributed by MDPI under the terms and conditions of the Creative Commons license CC BY-NC-ND.

Contents

About the Special Issue Editor vii

Preface to "Growth and Ecosystem Services of Urban Trees" ix

Astrid Moser-Reischl, Thomas Rötzer, Peter Biber, Matthias Ulbricht, Enno Uhl, Laiye Qu, Takayoshi Koike and Hans Pretzsch
Growth of *Abies sachalinensis* Along an Urban Gradient Affected by Environmental Pollution in Sapporo, Japan
Reprinted from: *Forests* 2019, 10, 707, doi:10.3390/f10080707 1

Rocco Pace, Peter Biber, Hans Pretzsch and Rüdiger Grote
Modeling Ecosystem Services for Park Trees: Sensitivity of i-Tree Eco Simulations to Light Exposure and Tree Species Classification
Reprinted from: *Forests* 2018, 9, 89, doi:10.3390/f9020089 20

Chi Zhang, Laura Myrtiá Faní Stratopoulos, Hans Pretzsch and Thomas Rötzer
How Do *Tilia cordata* Greenspire Trees Cope with Drought Stress Regarding Their Biomass Allocation and Ecosystem Services?
Reprinted from: *Forests* 2019, 10, 676, doi:10.3390/f10080676 38

Bertrand Festus Nero, Daniel Callo-Concha and Manfred Denich
Structure, Diversity, and Carbon Stocks of the Tree Community of Kumasi, Ghana
Reprinted from: *Forests* 2018, 9, 519, doi:10.3390/f9090519 52

Wan-Yu Liu and Ching Chuang
Preferences of Tourists for the Service Quality of Taichung Calligraphy Greenway in Taiwan
Reprinted from: *Forests* 2018, 9, 462, doi:10.3390/f9080462 69

Celina H. Stanley, Carola Helletsgruber and Angela Hof
Mutual Influences of Urban Microclimate and Urban Trees: An Investigation of Phenology and Cooling Capacity
Reprinted from: *Forests* 2019, 10, 533, doi:10.3390/f10070533 91

Susanne Jochner-Oette, Theresa Stitz, Johanna Jetschni and Paloma Cariñanos
The Influence of Individual-Specific Plant Parameters and Species Composition on the Allergenic Potential of Urban Green Spaces
Reprinted from: *Forests* 2018, 9, 284, doi:10.3390/f9060284 103

David Callow, Peter May and Denise M. Johnstone
Tree Vitality Assessment in Urban Landscapes
Reprinted from: *Forests* 2018, 9, 279, doi:10.3390/f9050279 117

Zhibin Ren, Xingyuan He, Haifeng Zheng and Hongxu Wei
Spatio-Temporal Patterns of Urban Forest Basal Area under China's Rapid Urban Expansion and Greening: Implications for Urban Green Infrastructure Management
Reprinted from: *Forests* 2018, 9, 272, doi:10.3390/f9050272 124

Gunwoo Kim and Paul Coseo
Urban Park Systems to Support Sustainability: The Role of Urban Park Systems in Hot Arid Urban Climates
Reprinted from: *Forests* 2018, 9, 439, doi:10.3390/f9070439 142

About the Special Issue Editor

Thomas Rötzer is Professor for Urban and Forest Ecosystem Modelling at the Technical University of Munich at the Research Department of Ecology and Ecosystem Management, Chair for Forest Growth and Yield Science. He is also Deputy Head of the Centre for Urban Ecology and Climate Adaptation (ww.zsk.de). He received a Ph.D. in horticultural sciences and was awarded his venia legendi for Forest Ecology and Modelling. His research and interests include tree and stand growth dynamics, process-based growth modeling, urban forestry, and the relationships between tree growth and ecosystem services, particularly under the view of climate change.

Preface to "Growth and Ecosystem Services of Urban Trees"

For the management of urban green areas, the great challenges in the future will be maintaining tree growth, enhancing tree vitality, and optimizing the provision of ecosystem services. The environmental conditions of cities worldwide will be changed substantially by increasing urbanization and through climate change. Urban green areas, and especially urban trees, are able to mitigate the negative effects of climate change by providing ecosystem services. They are carbon storage areas and, among others, they serve to mitigate the heat island effect, reduce rainwater runoff, filter pollutants, and provide shading and cooling effects. Additionally, they provide ecosystem services including recreation as well as health and quality of life effects. On the other hand, disservices like allergenic agents of trees or the release of biogenic volatile organic compounds can have negative effects or be harmful to human well-being. Quantitative values of these ecosystem services and disservices for different tree species depending on their size and environmental conditions are, however, hardly found in literature.

The services and disservices provided by an individual tree are closely linked with the tree species, the tree structure, the tree size and age, as well as with a tree's vitality and environment. The knowledge of urban tree growth in relation to these conditions is still poor. The physiological processes, the interactions, as well as the feedback reactions within the atmosphere–plant–soil system are scarcely understood. However, there is a great need of such knowledge for sustainable planning and management of urban green areas. Therefore, detailed knowledge about dimensional changes, growth rates, and ecosystem services of the most common urban tree species, depending on their age and on the environmental conditions, is necessary.

This Special Issue recognizes and deals with these research fields, which can be classified into three topics:

- assessing urban tree growth,

- deriving ecosystem services and disservices, and

- managing urban trees and their ecosystem services.

The 10 published articles in this book cover these topics. The articles of Moser-Reischl et al. (2019), Pace et al. (2018), and Zhang et al. (2019) can be assigned to the first theme, while the articles of Nero et al. (2018), Jochner-Oette et al. (2018), Liu and Chuang (2018), and Stanley et al. (2019) deal with ecosystem services and disservices issues. The articles of Callow et al. (2018), Ren et al. (2018), and Kim and Coseo (2018) fall in the category of managing urban trees and their ecosystem services. However, it is obvious that the publications of one topic are closely linked with the other topics.

Geographically, a wide range of climates and continents are covered. Four studies are located in Europe (i.e., in a temperate climate). Asia is represented by three studies: two of them were carried out in a tropical climate and one in a boreal climate. One study each was done in Africa, North America, and Australia.

A publication within the topic 'assessing urban tree growth' is the article of Moser-Reischl et al. (2019) who analyzed the growth of urban trees and trees of the rural surroundings for the city of Sapporo, Japan (i.e., in a boreal climate). They found higher growth rates for urban trees compared to rural trees in addition to an overall accelerated growth rate over time. Possible reasons are discussed. Pace et al. (2018) determined carbon sequestration, leaf area, and related ecosystem services of a park in the city of Munich. The uncertainty of emission simulations is discussed, and the importance to parameterize ecosystem functions for individual tree species is pronounced. Zhang et al. (2019) studied the responses of Tilia cordata in experimental water shortage and reported that this species reduced branch, stem, and coarse root biomass under heavy drought stress. Information on the fine and coarse root biomass development under drought is given as well as ecosystem services that are based on model simulations.

The study of Nero et al. (2018) can be assigned to the topic 'deriving ecosystem services and disservices'. The authors describe the structure of urban forests and the species composition for the city of Kumasi, Ghana, and they provide information on species richness and carbon storage potential. Liu and Chuang (2018) analyzed several recreational ecosystem services of a greenbelt in Taichung City, Taiwan. They recommend improving the cultural resources and the quality of recreational services. The paper of Stanley et al. (2019) is about urban tree growth and the regulation of ecosystem services along an urban heat island (UHI) gradient in the city of Salzburg, Austria. They demonstrate the influence of the UHI and of the tree characteristics on tree phenology, shading, and cooling capacity. Jochner-Oette et al. (2018) focus on ecosystem disservices (i.e., on the allergenicity of plants). They calculated an individual-specific allergenic potential index for park trees in the city of Eichstätt, Germany, investigated the effects of species composition, and gave recommendation for urban green planning.

The third topic is 'managing urban tree growth and its ecosystem services'. Callow et al. (2018) measured the drought stress of urban trees in Melbourne, Australia and analyzed climatic factors that are crucial for providing environmental services. They discuss methods for the assessment of long-term drought effects and other stressors on urban trees. Using thematic mapper imagery, Ren et al. (2018) analyzed spatio-temporal patterns of an urban forestry basal area index for the city of Changchun, China. They found that, over the studied period, the fragmentation of urban forests as well as the basal area of urban forests increased. Kim and Coseo (2018) estimated the ecosystem services of the urban park system of the city of Phoenix, USA. They valued the green infrastructure services of different urban vegetation types, which are fundamental for future urban green planning and management.

Altogether, the articles comprise important aspects of the urban green infrastructure, range over several climates, and include comprehensive information about urban tree growth and their ecosystem services.

Thomas Rötzer
Special Issue Editor

Article

Growth of *Abies sachalinensis* Along an Urban Gradient Affected by Environmental Pollution in Sapporo, Japan

Astrid Moser-Reischl [1,*], Thomas Rötzer [1], Peter Biber [1], Matthias Ulbricht [1], Enno Uhl [1], Laiye Qu [2,†], Takayoshi Koike [2,†] and Hans Pretzsch [1]

1 Forest Growth and Yield Science, School of Life Sciences, Weihenstephan, Technical University of Munich, Hans-Carl-von-Carlowitz-Platz 2, 85354 Freising, Germany
2 Graduate School of Agriculture, Hokkaido University, Sapporo 060-8589, Japan
* Correspondence: astrid.reischl@tum.de
† Present address: Urban and Regional Ecology, Research Center for Eco-Environmental Science, Beijing 100085, China and University of Chinese Academy of Sciences, Beijing 100049, China.

Received: 22 June 2019; Accepted: 16 August 2019; Published: 20 August 2019

Abstract: Urban tree growth is often affected by reduced water availability, higher temperatures, small and compacted planting pits, as well as high nutrient and pollution inputs. Despite these hindering growth conditions, recent studies found a surprisingly better growth of urban trees compared to trees at rural sites, and an enhanced growth of trees in recent times. We compared urban versus rural growing Sakhalin fir (*Abies sachalinensis* (F. Schmidt) Mast.) trees in Sapporo, northern Japan and analyzed the growth differences between growing sites and the effects of environmental pollution (NO_2, NO_X, SO_2 and O_X) on tree growth. Tree growth was assessed by a dendrochronological study across a gradient from urban to rural sites and related to high detailed environmental pollution data with mixed model approaches and regression analyses. A higher growth of urban trees compared to rural trees was found, along with an overall accelerated growth rate of *A. sachalinensis* trees over time. Moreover, environmental pollution seems to positively affect tree growth, though with the exception of oxides O_X which had strong negative correlations with growth. In conclusion, higher temperatures, changed soil nutrient status, higher risks of water-logging, increased oxide concentrations, as well as higher age negatively affected the growth of rural trees. The future growth of urban *A. sachalinensis* will provide more insights as to whether the results were induced by environmental pollution and climate or biased on a higher age of rural trees. Nevertheless, the results clearly indicate that environmental pollution, especially in terms of NO_2 and NO_X poses no threat to urban tree growth in Sapporo.

Keywords: air pollution; climate change implications; oxides; urbanity; tree growth

1. Introduction

The effects of air pollution and climate change on tree growth have been discussed ambiguously over the past decades. While several studies clearly link tree and forest damage with sensitivity to air and environmental pollution [1–6], some studies also named unfavorable climatic conditions with limited soil water availability [4,7], aggravated soil compaction, nutrient imbalances [8,9], and pests and disease infestation, as well as management errors as reasons for tree growth decline in forests [1,3,10,11]. On the other hand, several studies [12–16] found opposing positive effects of climate change conditions and increased environmental pollution on tree growth in forests. Presumed causes for the improved growth are higher nitrogen depositions, higher temperatures and higher CO_2-concentrations, together with a longer growing season and changes in forest management [12,17]. However, limited water availability might counteract these positive effects of climate change, shorten the growing season due to early leaf shed and reduce growth of trees [18]. Such contrasting findings have even been reported for single tree species. For example, Piovesan, et al. [19] found for *Fagus sylvatica* L. stands in Italy, a basal area increment (BAI) decrease of 15%–20% [19], while Pretzsch, Biber, Schütze, Uhl and Rötzer [16] reported a 30% increase of volume growth in Central Europe [16].

In contrast to forested sites, the effects of environmental pollution and climate on urban tree growth are less well understood. Several studies state a detrimental effect of environmental pollution on urban tree growth, phenology and vitality [20–24]. Studies reported changes in leaf anatomy and morphology, injury and reduced photosynthesis caused by heavy environmental pollution [24–28]. Kozlowski [29] stated foliar injury, higher mortality, reduced growth and yield, a reduction in shoot–produced compounds (carbohydrates) and stress to trees as the effects of environmental pollution [29]. Further, the impacts of climate change with warmer temperatures, higher maximum temperatures and less precipitation in summer will induce more stress on urban trees, possibly decreasing vitality and growth of less adapted species, and increase the risks of pests and disease. Therefore, the combined effect of environmental pollution and climate change's implications on trees should be regarded together [12]. The effects of environmental pollution and climate change on tree growth and vitality are highly important for urban trees, since the urban environment is overall a stressful, tough growing site for trees compared to forest sites [30]. This is due to conditions such as compacted, small planting pits, with reduced water and nutrient availability [31], root space [32] and aeration of root systems [33], as well as high temperatures [34] and mechanical injuries [35]. Additional negative influences, such as environmental pollution due to anthropogenic emissions might decrease growth and vitality to the limit of their sustainability [36]. Environmental pollution can weaken trees and open the door for insect infestations and pests [25,29]. Pollution is one of the major problems in urban environments for human health but also for tree vitality [36,37].

With an increasing urbanity along a gradient from the rural surroundings of a city to the inner-city centers, a reduced growth and vitality of trees might be expected due to the conditions outlined above. Surprisingly, the worldwide study of Pretzsch, Biber, Uhl, Dahlhausen, Schütze, Perkins, Rötzer, Caldentey, Koike, van Con, du Toit, Foster and Lefer [17], comparing rural and urban tree growth in several climate zones, found an enhanced growth of urban trees compared to the trees growing at the outskirts of many cities [17,38]. In that study, a total of 1383 urban trees were dendrochronologically sampled in ten metropolises worldwide, covering hemi to boreal (Sapporo, Japan; Prince George, Canada), temperate (Paris, France; Munich and Berlin, Germany), Mediterranean (Cape Town, South Africa; Santiago de Chile, Chile), and subtropical (Hanoi, Vietnam; Houston, MO, USA; Brisbane, Australia) climatic conditions [39,40]. The sampled trees of a defined species per city were selected from the city center to the suburban and rural areas and in all four primary directions from the city center. Dating back more than 100 years, the tree ring chronologies reflect the effect of global climate change and the urban heat island on urban tree growth worldwide. The study showed an increased growth rate of urban trees since the 1960s [17]. Moreover, across all cities and across the entire time span, urban trees grew more rapidly than those in the rural surroundings. This effect was most pronounced in the boreal climate zone. That was explained by higher temperatures and extended growing seasons in cities, as well as with increased CO_2-concentrations [17] similar to the study of Bytnerowicz, Omasa and Paoletti [12]. The urban heat island preempts the climate influence in general, but is most pronounced in the boreal climate [17]. In almost all investigated cities, except those in a temperate climate, the negative effects of the urban environment (e.g., reduction of photosynthesis by biogenic volatile organic compounds (BVOCs), fine dust and drought stress) are overcompensated by its benefits, such as an elongated growing period or fertilization due to emissions [15,41,42].

However, Guardans [2] found an increased climate change sensitivity of European beech and Norway spruce forest stands in boreal areas compared to other climatic zones, due to temperature and water stress, though reduced impacts of environmental pollution [2]. These contrasting results raise the question of how climate change and environmental pollution changes affect the growth of a coniferous urban tree species growing in the boreal climate zone. This study focuses on the growth of urban and rural *A. sachalinensis* MAST. trees in Sapporo, northern Japan, and the effects of urbanization, climate change and environmental pollution. This town rapidly increased its population from about one to two million in the past five decades. The growth of urban and rural trees was assessed by dendrochronology and related to climate and environmental pollution. The following research questions were stated:

1. What was the growth of *A. sachalinensis* in the past decades in the urban and rural areas of Sapporo, northern Japan?
2. Are the growth trends of *A. sachalinensis* similar to worldwide trends of urban tree growth?
3. Can differences in the growth of *A. sachalinensis* be found regarding the sampling sites?

2. Materials and Methods

2.1. Climate of Sapporo

Sapporo has about 2.0 million inhabitants and a size of 1121.12 km^2, the biggest city of Hokkaido island, northern Japan (43°4′ N, 141°21′ E ~ 43°3′43″ N, 141°21′15″E). Its climate is characterized as cold, without a dry season and hot summers [39,40] with an average annual temperature of 7.8 °C and a precipitation sum of 1130 mm (mean of 1980–2012). Over the year, the highest temperatures occur in July, August and September, with August as the hottest month (average of 20 °C). The coldest months are January and February with below 0 °C. The highest amount of precipitation also occurs in August and September (Figure 1, Japan Meteorological Agency).

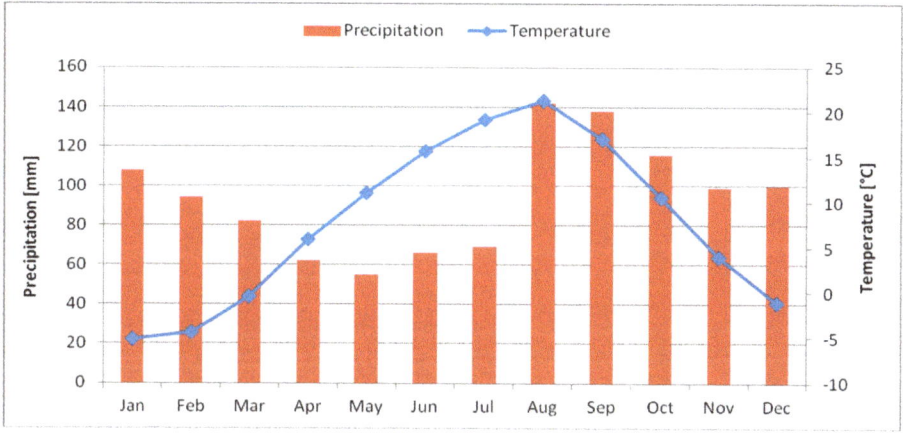

Figure 1. Average precipitation sums in mm and average annual temperature in °C from 1983 to 2012 in Sapporo, Japan provided by the Japan Meteorological Agency.

To further analyze the effect of climate on tree growth, we calculated the de Martonne-index [43] on the basis of monthly precipitation and monthly temperature from 1983 to 2012. The snow-free period is from mid-April to early November.

2.2. Environmental Pollution Data

Environmental pollution data (nitrogen dioxide NO_2, oxides O_X, sulfur dioxide SO_2 and nitric oxide NO_X) in Sapporo have been measured at several stations across the city (Table 1). Elevation of each monitoring site is about 50–200 m a.s.l. The highest concentrations of NO_2 were found at Kita-1-jyo and Tsukisamu-South, with the heavily trafficked region having a of mean concentration of 20.8 ppm. The highest O_X concentrations were found in Yamahana and Atsubetsu—the SE located suburb or close to reserved forests. The average concentrations of all stations was 26.9 ppm. For SO_2 we found mean concentrations of 4.4 ppm (highest concentrations at stations Sapporo Middle Part and West District with heavy traffic), while for NO_X the overall mean concentration was about 39.9 ppm (highest concentrations at stations Tsukisamu-South, near the reserve forest, and Kita-1-jyo, with heavy traffic). The average measured temperature of all stations across Sapporo was 8.6 °C. The warmest temperatures occurred at southeast Atsubetsu, in the east of Sapporo and at Yamahana, in the central south of Sapporo. Further, carbon monoxide CO was measured at the station Kita-1-jyo from 1986 to 2012, the mean value was 0.9 ppm. The values of NO_2 and O_X vary evenly around the mean, while SO_2, NO_X and temperature data are dominated by outliers, possibly induced by measurement errors or extreme conditions; e.g., at main roads.

Table 1. Environmental pollution (nitrogen dioxide: NO_2; oxides: O_X; sulfur dioxide: SO_2; and nitrous gases: NO_X) and climate data of several weather stations across Sapporo, Japan from 1983 to 2012, and variance from the mean value (+ higher than mean, – lower than mean). Added * to station name means suburb or closest to a green area (e.g., near an agriculture field, forested park or reserved forests).

Station	NO_2		O_X		SO_2		NO_X		Temperature	
Atsubetsu-SE*	21.9	+	29.4	+	3.2	–	38.7	–	9.1	+
EastDistrict	21.1	+	25.3	–	5.9	+	36.8	–	8.6	+
Fushimi-SW	18.5	–	26.0	–	-		31.0	–	8.6	+
Hassamu-NW*	16.9	–	28.0	+	4.2	–	26.2	–	8.6	+
Higashi 18-chome	25.4	+	-		-		54.6	+	-	
HigashiEast*	15.7	–	29.0	+	3.9	–	23.8	–	7.7	–
Kita-19-jyo	19.4	–	-		-		35.2	–	-	
Kita-1-jyo	35.0	+	-		-		85.0	+	-	
Kita-Shiroishi-E	18.2	–	28.1	+	1.0	–	28.6	–	8.8	+
MiddlePart	26.9	+	17.7	–	6.9	+	50.5	+	-	
Minami-14-jyo	22.4	+	-		-		47.7	+	-	
MinamiS	9.1	–	-		-		11.8	–	-	
ShinoroN*	13.1	–	28.5	+	3.9	–	19.9	–	8.0	–
Teine*	19.3	–	25.8	–	4.3	–	34.4	–	8.5	–
TsukisamuChuoS*	28.4	+	-		-		77.1	+	-	
WestDistrict	21.8	+	24.7	–	6.3	+	37.5	–	8.8	+
Yamahana	-		33.0	+	-		-		9.1	+
mean	20.8		26.9		4.4		39.9		8.6	

The change of the environmental pollution values in Sapporo over six time periods ranging from 1983 to 2012 is displayed in Figure 2. Although NO_X, NO_2 and SO_2 concentrations were decreasing, the O_X concentrations were increasing. The regression line showed a high coefficient of determination for NO_X, SO_2 and O_X ($R^2 > 0.65$) over time; however, for NO_2 it was lower ($R^2 < 0.25$). For NO_X and NO_2, there was a peak concentration recorded in the period 1998 to 2002. The overall highest concentrations of environmental pollution were found for NO_X, with values up to 50 ppm. The lowest concentrations were observed for SO_2.

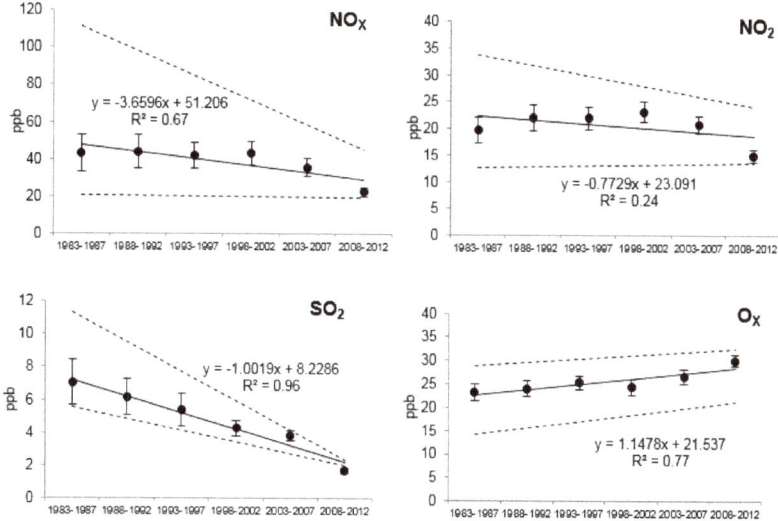

Figure 2. Minimum and maximum (dashed lines), as well as means with added standard error of environmental pollution data in six time periods (1983–1987, 1988–1992, 1993–1997, 1998–2002, 2003–2007 and 2008–2012) of several weather stations across the city of Sapporo. Given are the best fitting regression lines with regression coefficients and coefficients of determination R^2.

2.3. Sample Tree Species

The study focused on Sakhalin fir, *A. sachalinensis*, an evergreen conifer species originating of the Sakhalin islands and southern Kurils, Russia. The species also occurs in northern Hokkaido, Japan. It prefers moist climates with cool summers and mild winters, though it faces problems if exposed to waterlogged soils [44,45]. The shade tolerance of *A. sachalinensis* is very high and the growing rate low [46]. A screening experiment for 18 species native to Japan by Yamaguchi, et al. [47] showed, that most *Abies* species native to Japan are classified into the intermediate ozone sensitivity type (responses to AOT40; 16-30 ppm h), *A. sachalinensis* prefers slightly acidic soils with pHs around 5 [48].

2.4. Data Collection

Across the city of Sapporo, 109 *A. sachalinensis* individual trees were chosen for data sampling at six sites along a gradient from the city center to the suburbs of Sapporo and a forest area outside of the city (Figure 3).

The trees in sample plot 1 and plot 2 (Shirahata-yama 1 and Shirahata-yama 2) together with plot 6 (Misumai of Hokkaido University Forests) were all classified as rural (Table 2). The trees of plot 3 (Hokkaido University Nursery in Sapporo) were growing along street canyons and were therefore classified as urban. The trees of sample plots 4 and 5 (both Hitsujigaoka-7 site) were classified as suburban trees; however, for model development, were merged with the urban trees. The sample trees at the urban and suburban plots were typically trees growing in cities, standing along streets, in front of buildings and at squares. The trees at the forest sites were party planted, mostly due to esthetic reasons and not for timber production, since the wood of *A. sachalinensis* gains low prices in wood markets. The stem density at the rural plots spans from 2000 (plot 2), over 2200 (plot 6), to 2500 (plot 1) stems per hectare. To the best of our knowledge, the trees at plots 3 (urban), and 2 to 4 (rural to suburban) were not further managed. Trees of plots 1 (rural), as well as of plots 5 to 6 (suburban to rural) faced typhoon events or thinning. The soil nutrient status was slightly different across the sample plots. While urban and suburban plots had higher nitrogen (N) content and bulk densities than rural plots, the calcium (Ca) and magnesium (Mg) content was higher at rural plots.

The phosphorus (P) and potassium (K) content was not consistent over plot classification; however, highest concentrations were found for P at Shirahata-yama 2 and for K at Hitsujigaoka-7 2nd plots.

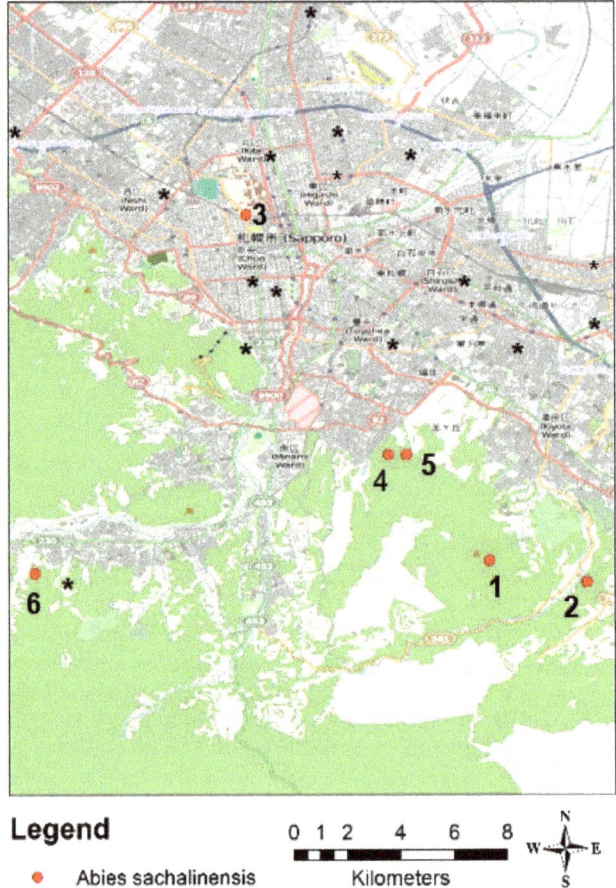

Figure 3. Sample sites of *Abies sachalinensis* within the city of Sapporo; the symbols * denotes approximately the locations of climate stations across Sapporo.

Table 2. Characteristics and soil chemical structure (nitrogen N, available phosphorus P content by the Bray ll method, potassium K, calcium Ca and magnesium Mg) at a depth of 10–20 cm, as well as bulk density of sample plots within Sapporo and its vicinity.

Plot	Name	Classification	N [mg/g]	P [mg/g]	K [µg/g]	Ca [mg/g]	Mg [mg/g]	Compaction [kg/m³]
1	Shirahata-yama 1	rural	2.02	4.23	148.00	-	-	0.50
2	Shirahata-yama 2	rural	2.11	6.81	182.00	-	-	0.50
3	Hokkaoido University Nursery	urban	3.07	1.89	152.00	2.44	0.23	0.55
4	Hitsujigaoka-7 1	suburban	3.02	1.69	345.00	1.88	0.22	0.52
5	Hitsujigaoka-7 2	suburban	3.78	2.33	360.00	-	-	0.60
6	Misumai	rural	2.98	1.98	288.00	3.31	0.45	0.59

Before increment core collection, data on tree structure and the site conditions were recorded, including diameter at breast height, 1.3 m (dbh), tree height (h), height to the crown base (cb), crown radius in four directions (N, E, S, W), tree position (coordinates and altitude), site condition, tree vitality, and open surface area of the unpaved area around the tree in four directions. Based on the recorded data, the average crown radius cr (Equation (1)) and crown diameter cd (cd = cr × 2), the crown projection area (CPA) (Equation (2)) and the crown volume (cv) (Equation (3)) were calculated.

$$\text{mean cr} = \sqrt{(r_N^2 + r_E^2 + r_S^2 + r_W^2)/4} \tag{1}$$

$$cpa = cr^2 \times \pi \tag{2}$$

$$cv = CPA \times \text{crown height} \tag{3}$$

The increment core collection was conducted at each tree. Two cores opposite to each other were extracted at a height of 1.3 m in northern and western directions with a 5 mm diameter increment corer (Haglöf Sweden AB; Långsele, Västernorrland, Sweden).

2.5. Core and Data Processing

The cores were processed by mounting on wooden boards with regard to the grain direction. Thereafter, the cores were sanded until the highest visibility of the cross-sectional area, and then polished with progressively finer sandpaper from grit size 180 up to 800. Annual tree-ring widths of the cores were measured with a digital positioning table with a resolution of 1/100 mm (Rinntech e.K., Heidelberg, Germany). Crossdating and synchronization of ring-width data were accomplished by the software TsapWin (Rinntech e.K., Heidelberg, Germany) using standard dendrochronological methods [49–51].

All following analyses were conducted with the package dplR of R [52]. The biological age trend (higher growth of younger trees) in the ring-width data was removed by a double detrending procedure applied to all series (modified negative exponential curves and cubic smoothing splines with 20 year rigidity, 50% wavelength cutoff). The resulting index series contained only year-to-year variability associated with fluctuations in climate [49,50] producing dimensionless ring-width indices (RWIs). In a final step, autocorrelation was removed by autoregressive (AR) models (maximum order of three) and the series were averaged using Tukey's biweight robust mean. This reduces bias caused by extreme values. Mean sensitivity was calculated as assessment of chronology quality; it depends on the year-to-year variability and was employed as a measure of variability. All further analyses of climate-growth correlations were conducted with the resulting chronologies. From the chronologies, the ages of the analyzed trees were derived. When the exact age of the tree was not clear (missing tree pith, among others), the age was back calculated based on the un-detrended average growth rate of the last ten years and the dbh of the tree.

2.6. Statistics

Data on previously measured tree structures (dbh, height and crown values) were tested for significant differences between groups of urbanity (urban versus rural versus suburban and urban versus rural). Since the assumptions of normality and homogeneity were not met, the Kruskal–Wallis test with pairwise testing for significance (pairwise Wilcox-test with Bonferroni–Holm p-value correction) was applied for three groups and the Kruskal–Wallis test for two groups. Statistics were done in R, version 3.3.3 [53].

2.7. Trend Analysis (Long-Term Trends)

Using the R package lme4, two linear mixed models of the following forms Equations (4) and (5) were developed to assess the influence of the time of age, growth (before 1960 and since 1960) and urbanity (urban-rural) on the annual basal area (response variable) derived from increment cores. To differentiate between the two growths-trend relevant periods (before 1960 and since 1960), we introduced the dummy variable recent, with 1 indicating each observation later than 1959 and 0 otherwise, which is in accordance with growth trends, since approximately the 1960s, that have been identified for forest trees.

$$\ln(ba_{ij}) = a_0 + a_1 \times time_{ij} + (b_0 + b_1 \times time_{ij}) \times \log(age_{ij}) + c_{ij} + \varepsilon_{ij}, \tag{4}$$

$$\ln(ba_{ij}) = a_0 + a_1 \times urb_{ij} + (b_0 + b_1 \times urb_{ij}) \times \log(age_{ij}) + c_{ij} + \varepsilon_{ij}, \tag{5}$$

In Equations (4) and (5) the basal area is the response variable for the jth of n_i observations in the ith of M groups or clusters, and $a_1, \ldots, a_n$ and $b_1, \ldots, b_n$ are the fixed effects with the a parameter's components of the intercept and the b parameter's components of the slope. When a_1 in Equation (4) differed significantly from 0, the age-basal area relationship before 1960 had a different intercept than that since 1960. In Equation (5), any differences in the intercepts would indicate that the intercept of urban trees was not the same as that for rural trees. The parameter b_1 in both equations have an analogous meaning to that of the slope. The c parameters are random effects, which are assumed to have a normal distribution. These random effects cover statistical dependencies, which are due to the nested data structure. The errors ε_{ij} are assumed to have an independent, identical distribution. With both models, we try to answer the research questions 1, 2 and 3 to derive the growth trend of A. sachalinensis in Sapporo in view of time and location.

3. Results

For the 109 measured A. sachalinensis trees, an average dbh of 34.4 cm with a mean age of 59 years was found. The mean tree height was 17.8 m and the average cd was 5.9 m. Moreover, an average cv of 1621 m^3 was found.

Table 3 displays the measured and calculated tree structures of A. sachalinensis in Sapporo classified by the sampling sites. The trees at rural plot 2 (Sharahata-yama 2) and rural plot 6 (Misumai) were the oldest with an average age of around 100 years. The according tree characteristics dbh, cd, cpa, crown length and cv were greatest at both plots as well. Youngest trees were found at suburban plots 4 and 5, Hitsujigaoka 1st and Hitsujigaoka 2nd, both in the Hitsukigaoka-7 site, with a mean age of around 35 years. Tree on those plots were, therefore, the smallest.

Table 3. Number of sampled trees per plot n with urbanity classification (urban, suburban, rural) and tree characteristics (age, dbh, tree height, crown start, crown diameter, crown projection area CPA, crown length and crown volume) of A. sachalinensis in Sapporo, Japan.

Plot	Classification	n	Age [a]	Dbh [cm]	Tree Height [m]	Crown Start [m]	Crown Diameter [m]	CPA [m^2]	Crown Length [m]	Crown Volume [m^3]
1	rural	20	47.3 ± 13.42	32.8 ± 6.8	17.9 ± 1.6	6.3 ± 2.4	5.5 ± 8.1	100.0 ± 207.7	11.6 ± 16.1	1161.5 ± 2382.3
2	rural	15	98.6 ± 23.2	49.1 ± 14.5	24.3 ± 3.6	7.3 ± 3.8	8.7 ± 2.4	253.6 ± 137.1	16.9 ± 4.7	4616.6 ± 3480.3
3	urban	18	40.4 ± 6.3	28.5 ± 4.6	15.2 ± 2.0	3.5 ± 1.5	4.9 ± 1.1	78.4 ± 35.9	11.7 ± 2.5	958.6 ± 529.1
4	suburban	21	32.2 ± 3.3	26.2 ± 2.9	15.3 ± 1.0	6.0 ± 2.0	5.2 ± 0.9	86.6 ± 33.4	9.3 ± 2.3	823.3 ± 415.9
5	suburban	20	35.3 ± 3.2	27.7 ± 2.8	16.0 ± 1.1	7.9 ± 1.4	4.5 ± 0.8	65.6 ± 19.9	8.1 ± 1.6	543.1 ± 219.5
6	rural	15	100.0 ± 64.1	41.9 ± 12.1	17.9 ± 2.4	6.4 ± 2.9	6.5 ± 1.5	138.1 ± 66.9	11.6 ± 2.9	1628.3 ± 1006.0

A further classification of A. sachalinensis to urban, suburban and rural growing trees illustrated that greatest tree structures were mostly found for rural trees (with exception of crown start). Urban and rural tree structures were significantly different, with suburban trees showing intermediate size (crown start), or were similar to the urban trees (dbh, tree height, cd, cpa and cv). Only the crown lengths of suburban trees were more similar to those of rural trees (Table 4).

Table 4. Tree structures (age, dbh, tree height, crown start, crown diameter, crown projection area cpa, crown length and crown volume) of *A. sachalinensis* of measured urban, suburban and rural trees with tested differences between categories. Different letters indicate significant differences found by one-way Kruskal–Wallis test and following a post-hoc test; n = number of sampled trees.

Site	n	Age [a]	Dbh [cm]	Tree Height [m]	Crown Start [M]	Crown Diameter [M]	CPA [m^2]	Crown Length [m]	Crown Volume [m^3]
urban	18	40.4 a ± 6.3	28.5 a ± 4.6	15.2 a ± 2.0	3.5 a ± 1.5	4.9 a ± 1.1	78.4 a ± 35.9	11.7 a ± 2.5	958.6 a ± 529.1
suburban	43	34.4 a ± 2.2	26.5 a ± 2.9	15.6 a ± 1.1	7.0 ab ± 2.0	4.8 a ± 0.9	76.1 a ± 29.2	8.6 b ± 2.1	678.5 b ± 359.7
rural	50	82.0 b ± 30.03	40.4 b ± 12.9	19.8 b ± 3.9	6.6 b ± 3.8	6.7 b ± 2.2	157.5 b ± 86.6	13.2 a ± 4.1	2338.1 b ± 1874.2

3.1. Short-Term Growth Trends

When analyzing the overall growth trends of *A. sachalinensis* in Sapporo, a strong age-trend could not be found (Figure 4a). At younger ages, *A. sachalinensis* did not show a better radius growth than older trees displayed. However, with the exception of 2011, a decrease in growth can be observed during the last ten years. After double-detrending, these trends cannot be found in Figure 4b, highlighting a more varying indexed growth earlier in life and during the past years, and a very uniform growth from 1984 to 2008. Overall, the basal area of *A. sachalinensis* shows a steady increase, with a drop in 1956, 1995 and 2011 (Figure 4c).

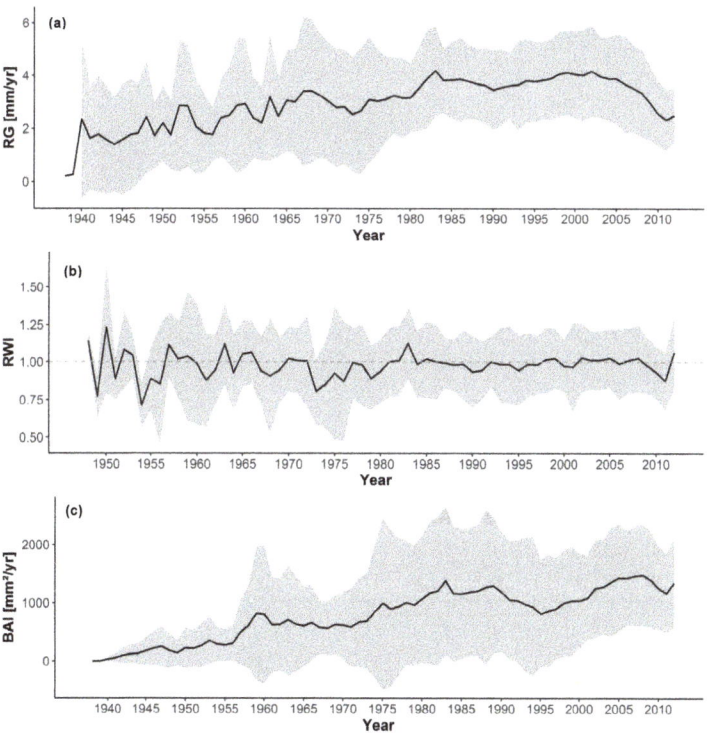

Figure 4. Radius growth, RG (a), indexed growth values, ring-width indices (RWI) (b), and basal area increments, BAI (c), of all sampled *A. sachalinensis* trees in Sapporo.

When looking in more detail at the growth of *A. sachalinensis* in Sapporo, analyzing the growth chronologies of trees at urban, suburban and rural settings, distinct trends can be found (Figure 5). The raw radial growth curves of rural trees displayed a far higher age of rural trees than of urban and suburban trees (Figure 5a). Urban and suburban chronologies started the earliest, in 1985. When regarding their growth, rural trees showed lowest stem growth rates, while urban and suburban trees indicated clearly higher values. For all chronologies, a decline was found for the last ten years, which was most pronounced for suburban and urban trees. After the detrending process and after the age-trend removal (Figure 5b), the growth chronologies appear more similar. However, common positive or negative pointer years for all three chronologies could not be found. The decline in the past years was observed as well, albeit a steep increase in growth was found in 2011. The basal area increment (Figure 5c) indicates a higher level of BAI of rural trees due to their higher age and longer growing period. Both urban and suburban trees had a marked increase in BAI since their planting, exceeding rural trees between 2005 and 2008.

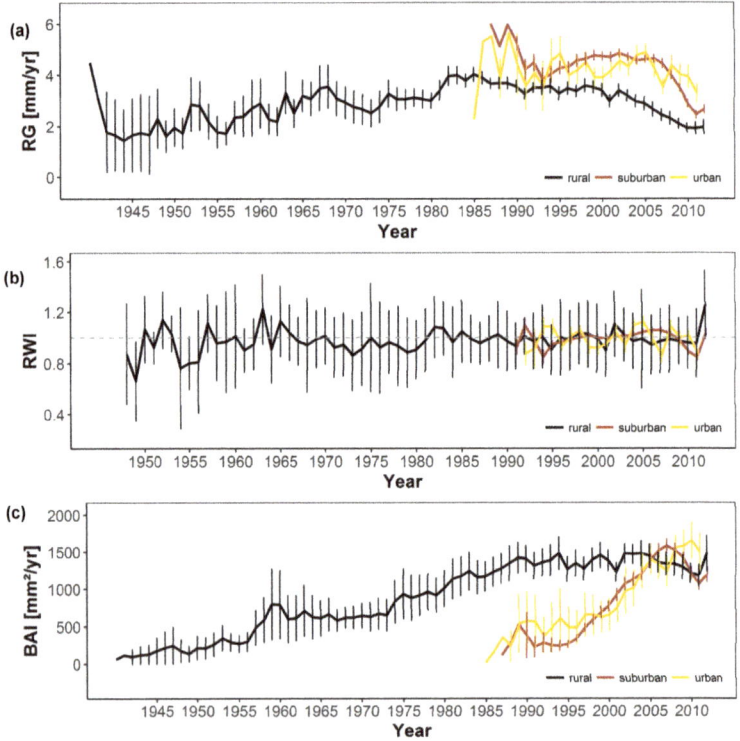

Figure 5. Radius growth, RG (**a**), indexed growth values RWI (**b**) and basal area increments BAI (**c**) with standard error bars for each value of *A. sachalinensis* trees sampled in rural, suburban and urban areas of Sapporo.

The statistical characteristics of the displayed chronologies can be found in Table 5. While the highest growth rate was found for urban and suburban trees, rural trees were growing on average the least. Mean sensitivity was 0.23, with highest for the urban chronology and lowest for the suburban chronology (0.19). Autocorrelation was found highest for rural trees, at 0.68 and least for urban trees, at 0.33.

Table 5. Mean radius growth rate [mm/yr], mean sensitivity and autocorrelation (autoregressive, AR) of urban, suburban and rural chronologies, as well as for the overall mean chronology of A. sachalinensis in Sapporo, Japan for their total period of growth.

Chronology	Growth Rate	Mean Sens.	AR
Urban	4.22	0.23	0.33
Suburban	4.21	0.19	0.51
Rural	3.46	0.21	0.68
overall	3.88	0.20	0.56

3.2. Long-Term Growth Trends

Tree growth in Sapporo has shown an enhanced growth for the past decades compared with the growth before 1960. This is in accordance with the overall global trend of urban trees found by [17] (Figure 6a). Young *A. sachalinensis* grew similarly, regardless of their planting time; however, after reaching an age of around 50 years, trees planted in more recent decades showed an enhanced growth, also passing the overall worldwide average. Urban *A. sachalinensis* trees illustrated a far better growth compared to rural trees, in particular after an age of 60 years, urban tree growth exceeded rural trees' growth by far (Figure 6b). In comparison to the worldwide results on urban and rural tree growth by Pretzsch, Biber, Uhl, Dahlhausen, Schütze, Perkins, Rötzer, Caldentey, Koike, van Con, du Toit, Foster and Lefer [17], *A. sachalinensis* rural trees showed a lower average growth rate. Younger trees were growing less than the worldwide average, though, at the old age of more than 100 years, *A. sachalinensis* exceeded the growth rate of other studies trees. Urban trees had a similar growth rate to those trees measured worldwide until an age of 50 years, and then an extremely high growing rate was found for *A. sachalinensis* in Sapporo.

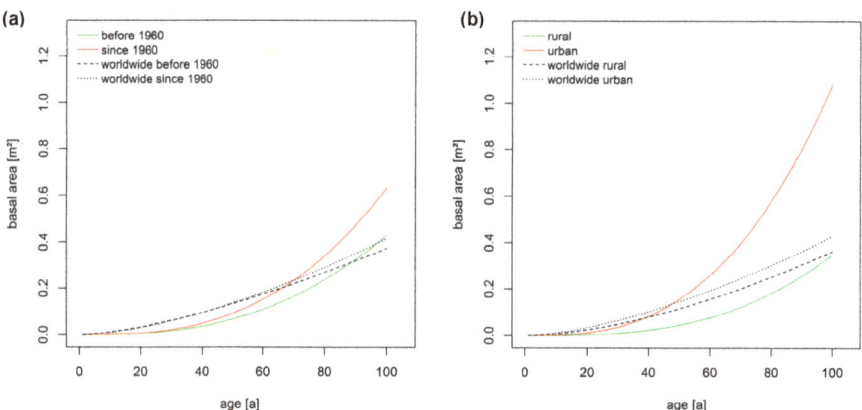

Figure 6. The growth in size in terms of basal area growth of *A. sachalinensis* in Sapporo and for worldwide urban trees [17] for (**a**) the period before 1960 compared with the period since 1960 and (**b**) urban trees compared with rural growing trees.

The described trends of Figure 6a,b can also be found by using a mixed model (Tables 6 and 7). While age had an overall significantly positive effect on growth of *A. sachalinensis*, the time of growth (before or since 1960) acted as negative driver, though it was not significant. In combination with age, a significant, positive interaction was found again. Therefore, the older trees were, the better was the tree growth since 1960. A similarly positive effect for age on tree growth as a single term was found when looking at the influence of urbanity. Urbanity had a highly positive influence on growth, indicating a better growth of urban *A. sachalinensis* compared to rural trees. In combination, however, urban trees showed a more enhanced growth with higher age than rural trees.

Table 6. Results of the linear mixed model on the annual basal area increment (mm^2 yr^{-1}) of all analyzed trees (response variable) with the individual tree coded as random effect, and the fixed effects were log (age), time of growth (period before 1960 and period since 1960) and their interaction.

Parameter	Fixed Effect	Value ± SE	p
a	Intercept	−13.14 ± 0.66	<0.001
Log (age_{ij})	Log (age)	2.67 ± 0.04	<0.001
$time_{ij}$	Time of growth	−0.15 ± 0.14	0.26
$b \times time_{ij} \times \log(age_{ij})$	Log (age) × Time of growth	0.12 ± 0.04	<0.001
Random effect d_{ij}		1.07	-
ε		0.32	-

Levels of Time of growth: 2 (Before 1960 and Since 1960).

Table 7. Results of the linear mixed model on the annual basal area increment (mm^2 yr^{-1}) of all analyzed trees (response variable) with the individual tree code as random effect. The fixed effects are log (age), the grade of urbanity (urban and rural) and their interaction.

Parameter	Fixed Effect	Value ± SE	p
a	Intercept	−14.58 ± 0.15	<0.001
Log (age_{ij})	Log (age)	2.94 ± 0.02	<0.001
urb_{ij}	Urbanity	1.79 ± 0.20	<0.001
$b \times urb_{ij} \times \log(age_{ij})$	Log (age) × Urbanity	−0.14 ± 0.02	<0.001
Random effect d_{ij}		0.82	-
ε		0.32	-

Levels of Urbanity: 2 (Rural and Urban).

3.3. Relationships of Growth with Environmental Pollution

The growth was also related to the measured environmental pollution and climate data of Sapporo (Figure 7).

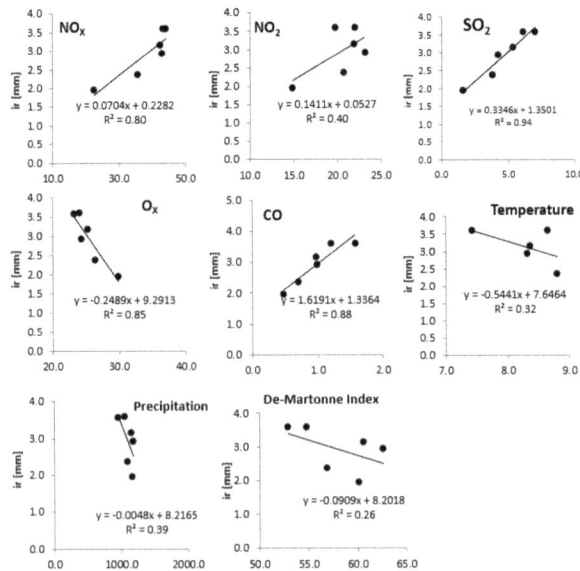

Figure 7. Relationships of mean environmental pollution (NO$_X$, NO$_2$, SO$_2$, O$_X$, CO) and mean climate data (temperature, precipitation, de Martonne-Index) with average radial growth of *A. sachalinensis* in Sapporo, Japan in five-year time steps from 1983 to 2012 and an added regression line with regression coefficients and adjusted R^2.

Radius increment, concentrations of environmental pollutions and climate data were averaged in six periods of five years. Over the analyzed periods, strong correlations of growth with NO_X, SO_2, O_X and CO were found ($R^2 > 0.8$). However, weak relationships were found between growth and NO_2, temperature, precipitation and the de Martonne-index ($R^2 < 0.4$). When looking at the strengths of the relationships, a strong positive effect of higher environmental pollution on tree growth in Sapporo was derived for NO_X, NO_2, SO_2 and CO. An increase in concentrations of these gases was strongly related to an increase in growth. On the other hand, for the O_X concentration and the climate data temperature, precipitation and the de Martonne-index, a negative relationship with growth was observed. A higher concentration of O_X as well as higher temperatures, more precipitation and a higher de Martonne-index was correlated with decreased growth.

4. Discussion

The studied *A. sachalinensis* trees in Sapporo, Japan were, across all sampling sites, an average age of 59 years old. This age is comparatively high for an urban tree species, since studies such as Nowak, et al. [54] and Skiera and Moll [55], stated a very low life span of 15 years on average and a high mortality rate of 6.6% per year. However, the most frequently planted trees in cities are deciduous tree species, illustrating a very high annual mortality rate of 12.6% for *Acer negundo* L., 7.4% for *Robinia pseudoacacia* L. and 6.8% for *Fraxinus pennsylvanica* Marsh., compared to results for coniferous species (3.2% for *Picea abies* (L.) Karst. and 0% for *Pinus strobus* L.) [54]. In a Japanese forest stand, a competition driven mortality rate of 1.7% per year was found for *A. sachalinensis*, with an overall age range of 33 to 187 years and a mean age of 104 years in the Nakagawa Experiment Forest of Hokkaido University, located in the northern part of the island [56]. Moreover, a maximum age of 249 years for canopy trees of *A. sachalinensis* was derived [56]. This is in line with the maximum age we found in our study, as a maximum age of 250 years was found for trees sampled at the rural Misumai plot. The oldest trees were found at the rural plots Shirahata-yama 2 and Misumai. The trees at suburban and urban plots were markedly younger. These plots were in the rural outskirts of the city and have been established earlier than the other urban and suburban sites. The average age of all sampled trees across Sapporo was younger compared to the study of Hiura, Sano and Konno [56]; however, when excluding the fairly young urban and suburban trees, an average age of 82 years for the trees at the forested sites can be derived. When only looking at plots Shirahata-yama 2 and Misumai, the average age of the trees was 100 years, which is very similar to the study about *A. sachalinensis* growing in a Japanese forest [56]. All in all, the found growth rates and ages seem to be in line with the results of other studies on coniferous species, urban trees and *A. sachalinensis*.

Tree age is correlated with tree structures like diameter at breast height (dbh), tree height and crown parameters, such as diameter, height, projection area and volume [57], and the older trees at rural plots Shirahata-yama 2 and Misumai had significantly greater dbh, height and other crown parameters than the trees growing at suburban and urban plots. Due to their young age, though, urban and suburban trees show a high growth rate. While the trees sampled at the forest-sites are more than double as old as the trees at suburban and urban plots, the difference in dbh is only 14 cm and 12 cm respectively. Overall, the height growth of *A. sachalinensis* in Sapporo is faster compared to trees measured in a northern Japanese forest stand [56]; however, urban trees often can reach greater tree sizes than forest trees due to more unobstructed growth and less competition [58].

The average raw radius growth rate of *A. sachalinensis* in Sapporo, northern Japan is with 4.2 mm/year, in line with other studies analyzing the growth of this species, such as Hiura, Sano and Konno [56], who found a growth rate of 1.0–7.0 mm/year, and Umeki [59] describing 3.1 mm per year, both studying *A. sachalinensis* at forest-sites in Japan. In urban areas, the growth rates of coniferous species have seldom been measured. Peper, et al. [60] provided data on the dbh increments between the ages 15 and 30 years for three coniferous species in Santa Monica, California, US. The transformed data indicate radius increments of 5.7 mm per year for *Pinus canariensis* C. Sm., 4.7 mm for *Ficus microcarpa* L. f. and 8.4 mm for *Cedrus deodara* (D. Don.) G. Don., which all are higher than those found for

A. sachalinensis in Japan. When accounting for the similar age, the growth rate *F. microcarpa* is comparable to the radius increments of urban and suburban trees of *A. sachalinensis*. The highest growth was found for urban and suburban *A. sachalinensis* trees, while trees at rural plots grew least. This might be induced by the younger age of urban and suburban trees.

For all further analysis, the age difference was accounted for in model development by including the age of the trees or by an age detrending of the raw tree ring widths. A similar year-to-year variability (mean sensitivity) was found for the urban, suburban and rural chronologies, albeit trees at the urban sites showed higher mean sensitivity. This seems logical, since trees growing in extreme environments often display stronger sensitivity values than trees of sites with less stress and better water supply [61]. The higher autocorrelation of rural trees indicated, that radial growth was strongly influenced by the conditions of the preceding year, due to preformed shoot primordia in this species [62]. The lower autocorrelation of urban trees might be caused by the different conditions at the urban sites compared to forest sites. The urban environment seems to affect trees positively, while rural trees are more often faced with stressful conditions, such as possibly less nutrient inputs and higher ozone concentrations over 60~80 ppb (nmol mo^{-1}). Information on the soil nutrient status showed, that the soil at rural plots has higher phosphorus and magnesium concentrations; however, the nitrogen content is lower. Moreover, the soil compaction is slightly less at rural sites.

The radius and BAI chronology of urban trees also indicated a growth decline in the past decades, though the steep increase in growth during the past year before sample taking. In total, the chronology of the urban trees showed a decreased variability and a higher stability over the past decades. The chronologies of urban and suburban trees illustrated high radius growth rates and strong increases in BAI. Though urban and suburban chronologies highlighted an opposing pattern to the rural trees in past decades (decrease in radius increment and BAI), this is in particular visible for the suburban chronology.

Trees of *A. sachalinensis* in Sapporo showed an enhanced growth rate during the most recent decades compared to former times (before 1960), as well as a better growth of urban trees compared to rural trees. These findings are in line with the trends found for *R. pseudoacacia* in Santiago de Chile [63] and *Quercus nigra* Mill. in Houston, Texas [42]. However, results of *A. sachalinensis* might be biased by the strong age and growth differences of urban and rural trees, since no sampled urban tree was older than 47 years (planted around 1970–1975). Trees of the category "before 1960" were only rural trees, which illustrated a lower growth rate than urban trees. The different age of the trees was included in the model performance, though the great difference might still bias the results slightly.

In total, the results of tree growth in Sapporo follow the overall worldwide trend [17] found for trees in metropolises, which describes an enhanced growth of urban trees, possibly induced by a longer growing period, higher CO_2-concentrations and warmer temperatures compared to rural sites [15,41,64]. The accelerated growth of urban trees was most pronounced for the boreal zone. Urban trees of this zone, possibly gain most by higher temperatures, by higher nutrient inputs and by prolonged growing periods, especially when further water availability is sufficient [17]. An extended growing period by climate change, though, will not affect *A. sachalinensis*, as much as it might affect other deciduous species. Since *A. sachalinensis* is a coniferous species, it can use photosynthesis throughout the whole year, and much longer than deciduous species anyway. Guardans [2], on the other hand, described an increased climate change and environmental pollution sensitivity of trees in the boreal area due to temperature and water stress. This was also found by Rötzer, et al. [65] for urban trees in temperate cities. Due to the relative high precipitation amounts in Sapporo with more than 1000 mm per year, though, this pollution sensitivity to temperature and water stress might not be true for *A. sachalinensis*. Moreover, correlating radial growth of this species with precipitation and a climate index, the de Martonne-index revealed a negative effect of increased precipitation on growth. Similarly, higher temperatures were negatively correlated with growth as well. Together with a negative influence of a higher de Martonne-index, it seems that possibly, a high water-availability and warmer temperatures have surprisingly negative effects on growth, albeit correlation coefficients were low ($R^2 < 0.4$). As this species is adapted to cold boreal climate zones, the negative relationship with temperature seems likely. Moreover, Temperate Plants Database [66] and Chittendon [44] described *A. sachalinensis* as a species growing well on moist soils, but it is, however, sensitive to water logging. If soils are not well-drained, growth and vitality declines might follow. Since high run-off rates are typical for urban soils, a stronger waterlogging effect during the snow melting season in rural areas could occur.

In summary, neither a prolonged growing season, nor higher temperatures induced the better growth of urban trees. A further correlation with environmental pollution revealed that NO_X, SO_2, CO and NO_2 had strong positive effects on urban tree growth. Only with O_X, a negative correlation was found. Related to Figure 7, SO_2 is originated from diesel cars with the use of relatively low-quality gas, and its trend is similar to the emission of NO gas. These (SO_X and NO) decreases were accompanied by an increase in O_3. This O_3 is partly provided from transboundary O_3 from windward regions [47]. Firs growing in Japan, such as *Abies* sp., were classified as mid-tolerant tree species in terms of elevated O_3, among 25 tree seedlings, through a big size screening text with Open Top Chambers [46,47]. The fir especially, is sensitive at the needle flushing time of mid-April to early May, when O_3 concentration usually reaches 50–70 ppb. Several studies reported a negative effect of ozone concentrations to tree growth [47,67–70]. Due to photosynthesis inhibition found for street trees [71], later bud break times and reduced leafs per bud were found for *F. crenata* Blume [47,72]. Moreover, these findings overlap with the results of Gregg, et al. [73], who described a strong negative effect of ozone concentration on the growth of clonal poplar. A reduced ozone exposure at urban sites was stated as reason for better growth of urban plants compared to higher ozone concentration and reduced growth of rural plants [73].

However, a decrease in environmental pollution in Sapporo was found during past years, though O_X was increasing. Therefore, a monitoring of *A. sachalinensis* in future will provide further insights. The bias by the age differences of the urban trees in urban and rural surroundings might then be excluded, and the influence of atmospheric concentrations and risks of waterlogging could be analyzed in more detail.

5. Conclusions

A. sachalinensis trees in Sapporo, northern Japan showed distinct growth patterns related to their growing site. Very old rural trees had lower growth rates than younger urban trees. Moreover, an enhanced growth of trees during past years was found. Growth relationships with environmental pollutions were mainly positive. Only O_X seemed to be a major cause of reduced rural tree growth, possibly together with higher tendencies of waterlogged soils, though this was not analyzed in our study. In total, environmental pollution and warmer temperatures, as well as further growing conditions at urban sites (e.g., small planting pits and compacted soils) did not affect tree growth as negatively as could be expected.

Author Contributions: H.P., T.R, E.U. and P.B. conceptualized the experiment. T.R., M.U., L.Q. and T.K. collected the data and provided help with obtaining permission. P.B. invented the statistical analysis. A.M.-R. analyzed the data and wrote the manuscript. All authors provided editorial advice and participated in the review process.

Funding: This research was funded by the AUDI Environmental Foundation for funding this study (project 5101954: "Reaktionskinetik von Bäumen unter Klimaveränderungen"—"Reaction kinetics of trees under climate change").

Acknowledgments: All contributors thank the municipal authority of Sapporo and Field Science Center for Northern Biosphere of Hokkaido University, Japan, for supporting the search for the trees and the permission to measure and core the trees.

Conflicts of Interest: The authors declare no conflict of interest.

References

1. De Bauer, L.I.; Krupa, S.V. The Valley of Mexico: Summary of Observational Studies on its Air Quality and Effects on Vegetation. *Environ. Pollut.* **1990**, *65*, 109–118. [CrossRef]
2. Guardans, R. Estimation of climate change influence on the sensitivity of trees in Europe to air pollution concentrations. *Environ. Sci. Policy* **2002**, *5*, 319–333. [CrossRef]
3. Juknys, R.; Vencloviene, J.; Stravinskiene, V.; Augustaitis, A.; Bartkevicius, E. Scots pine (*Pinus sylvestris* L.) growth and condition in a polluted environment: From decline to recovery. *Environ. Pollut.* **2003**, *125*, 205–212. [CrossRef]
4. Kint, V.; Aertsen, W.; Campioli, M.; Vansteenkiste, D.; Delcloo, A.; Muys, B. Radial growth change of temperate tree species in response to altered regional climate and air quality in the period 1901–2008. *Clim. Chang.* **2012**, *115*, 343–363. [CrossRef]
5. Legge, A.H.; Jager, H.J.; Krupa, S.V. Sulfur dioxide. In *Recognition of Air Pollution Injury to Vegetation: A Pictorial Atlas*; Flagler, R., Ed.; Air and Waste Management Association: Pittsburgh, PA, USA, 1999; pp. 3/1–3/42.
6. Lorenz, M.; Clarke, N.; Paoletti, E.; Bytnerowicz, A.; Grulke, N.; Lukina, N.; Sase, H.; Staelens, J. Air pollution impacts on forests in changing climate. In *Forest and Society–Responding to Global Drivers of Change*; IUFRO World Series; Mery, G.E.A., Ed.; International Union of Forest Research Organizations: Vienna, Austria, 2010; pp. 55–74.
7. Scharnweber, T.; Manthey, M.; Criegee, C.; Bauwe, A.; Schroder, C.; Wilmking, M. Drought matters—Declining precipitation influences growth of Fagus sylvatica L. and Quercus robur L. in north-eastern Germany. *For. Ecol. Manag.* **2011**, *262*, 947–961. [CrossRef]
8. Dittmar, C.; Zech, W.; Elling, W. Growth variations of common beech (*Fagus sylvatica* L.) under different climatic and environmental conditions in Europe—a dendroecological study. *For. Ecol. Manag.* **2003**, *173*, 63–78. [CrossRef]
9. Penninckx, V.; Meerts, P.; Herbauts, J.; Gruber, W. Ring width and element concentrations in beech (*Fagus sylvatica* L.) from a periurban forest in central Belgium. *For. Ecol. Manag.* **1999**, *113*, 23–33. [CrossRef]
10. Chappelka, A.H.; Grulke, N. Disruption of the 'disease triangle' by chemical and physical environmental change. *Plant Biol.* **2016**, *18*, 5–12. [CrossRef]
11. Innes, J.L. *Forest Health: Its Assessment and Status*; CAB International Publishers: Oxon, UK, 1993.
12. Bytnerowicz, A.; Omasa, K.; Paoletti, E. Integrated effects of air pollution and climate change on forests: A northern hemisphere perspective. *Environ. Pollut.* **2007**, *147*, 438–445. [CrossRef]
13. Spiecker, H.; Mielikäinen, K.; Kölh, M.; Skovsgaard, J.P. *Growth Trends in European Forests: Studies from 12 Countries*; Springer: Berlin, Germany, 1996.
14. Fang, J.; Kato, T.; Guo, Z.; Yang, Y.; Hu, H.; Shen, H.; Zhao, X.; Kisimoto-Mo, A.W.; Tang, Y.; Houghton, R.A. Evidence for environmentally enhanced forest growth. *Proc. Natl. Acad. Sci. USA* **2014**, *111*, 9527–9532. [CrossRef]
15. Churkina, G.; Zaehle, S.; Hughes, J.; Viovy, N.; Chen, Y.; Jung, M.; Heumann, B.W.; Ramankutty, N.; Heimann, M.; Jones, C. Interactions between nitrogen deposition, land cover conversion, and climate change determine the contemporary carbon balance of Europe. *Biogeosciences* **2010**, *7*, 2749–2764. [CrossRef]
16. Pretzsch, H.; Biber, P.; Schütze, G.; Uhl, E.; Rötzer, T. Forest stand growth dynamics in Central Europe have accelerated since 1870. *Nat. Commun.* **2014**, *5* (Suppl. 1), 4967. [CrossRef] [PubMed]
17. Pretzsch, H.; Biber, P.; Uhl, E.; Dahlhausen, J.; Schütze, G.; Perkins, D.; Rötzer, T.; Caldentey, J.; Koike, T.; van Con, T.; et al. Climate change accelerates growth of urban trees in metropolises worldwide. *Sci. Rep.* **2017**, *7*, 1–10. [CrossRef] [PubMed]
18. Rötzer, T.; Biber, P.; Moser, A.; Schäfer, C.; Pretzsch, H. Stem and root diameter growth of European beech and Norway spruce under extreme drought. *For. Ecol. Manag.* **2017**, *406*, 184–195. [CrossRef]
19. Piovesan, G.; Biondi, F.; Di Filippo, A.; Alessandrini, A.; Maugeri, M. Drought-driven growth reduction in old beech (*Fagus Sylvatica* L.) forests of the central Apennines, Italy. *Glob. Chang. Biol.* **2008**, *14*, 1265–1281. [CrossRef]
20. Alaimo, M.G.; Lipani, B.; Lombardo, M.G.; Orecchio, S.; Turano, M.; Melati, M.R. The mapping of stress in the predominant plants in the city of Palermo by lead dosage. *Aerobiología* **2000**, *16*, 47–54. [CrossRef]

21. Baycu, G.; Tolunay, D.; Özden, H.; Günebakan, S. Ecophysiological and seasonal variations in Cd, Pb, Zn, and Ni concentrations in the leaves of urban deciduous trees in Istanbul. *Environ. Pollut.* **2006**, *143*, 545–554. [CrossRef]
22. Bhatti, G.H.; Iqbal, M.Z. Investigations into the effect of automobile exhausts on the phenology, periodicity and productivity of some roadside trees. *Acta Soc. Bot. Pol.* **1988**, *57*, 395–399. [CrossRef]
23. Koike, T.; Watanabe, M.; Hoshika, Y.; Kitao, M.; Matumura, H.; Funada, R.; Izuta, T. Effects of ozone and forest ecosystems in East and Southeast Asia. *Dev. Environ. Sci.* **2013**, *13*, 371–390.
24. Pourkhabbaz, A.; Rastin, N.; Olbrich, A.; Langenfeld-Heyser, E.; Polle, A. Influence of Environmental Pollution on Leaf Properties of Urban Plane Trees, *Platanus orientalis* L. *Bull. Environ. Contam. Toxicol.* **2010**, *85*, 251–255. [CrossRef]
25. Izuta, T. *Air Pollution Impacts on Plants in East Asia*; Springer: Tokyo, Japan, 2017.
26. Joshi, P.C.; Abhishek, S. Physiological responses of some tree species under roadside automobile pollution stress around city of Haridwar, India. *Environmentalist* **2007**, *27*, 365–374. [CrossRef]
27. Pandey, J.; Agrawal, M. Evaluation of air pollution phytotoxicity in a seasonally dry tropical urban environment using three woody perennials. *New Phytol.* **1994**, *126*, 53–61. [CrossRef]
28. Williams, R.J.H.; Lloyd, M.M.; Ricks, G.R. Effects of atmospheric pollution on deciduous woodland I: Some effects on leaves of *Quercus petraea* (Mattuschka) Leibl. *Environ. Pollut.* **1971**, *2*, 57–68. [CrossRef]
29. Kozlowski, T.T. The impact of environmental pollution on shade trees. *J. Arboric.* **1986**, *12*, 29–37.
30. Moser, A.; Rötzer, T.; Pauleit, S.; Pretzsch, H. Structure and ecosystem services of small-leaved lime (*Tilia cordata* Mill.) and black locust (*Robinia pseudoacacia* L.) in urban environments. *Urban For. Urban Green.* **2015**, *14*, 1110–1121. [CrossRef]
31. Morgenroth, J.; Buchan, G.D. Soil moisture and aeration beneath pervious and impervious pavements. *Arboric. Urban For.* **2009**, *35*, 135–141.
32. Bühler, O.; Kristoffersen, P.; Larsen, S.U. Growth of Street Trees in Copenhagen With Emphasis on the Effect of Different Establishment Concepts. *Arboric. Urban For.* **2007**, *5*, 330–337.
33. Rahman, M.A.; Stringer, P.; Ennos, A.R. Effect of Pit Design and Soil Composition on Performance of *Pyrus calleryana* Street Trees in the Establishment Period. *Arboric. Urban For.* **2013**, *39*, 256–266.
34. Akbari, H.; Pomerantz, M.; Taha, H. Cool surfaces and shade trees to reduce energy use and improve air quality in urban areas. *Sol. Energy* **2001**, *70*, 295–310. [CrossRef]
35. Beatty, R.A.; Heckman, C.T. Survey of urban tree programs in the United States. *Urban Ecol.* **1981**, *5*, 81–102. [CrossRef]
36. Rodríguez Martín, J.A.; De Arana, C.; Ramos-Miras, J.J.; Gil, C.; Boluda, R. Impact of 70 years urban growth associated with heavy metal pollution. *Environ. Pollut.* **2015**, *196*, 156–163. [CrossRef] [PubMed]
37. Escobedo, F.J.; Nowak, D.J. Spatial heterogeneity and air pollution removal by an urban forest. *Landsc. Urban Plan.* **2009**, *90*, 102–110. [CrossRef]
38. Ding, H.; Pretzsch, H.; Schütze, G.; Rötzer, T. Size dependency of tree growth response to drought among Norway spruce and European beech individuals in monospecific and mixed-species stands. *Plant Biol.* **2017**. [CrossRef] [PubMed]
39. Köppen, W.; Geiger, G. *Handbuch der Klimatologie*; Gebrüder Borntraeger: Berlin, Germany, 1930–1939.
40. Peel, M.C.; Finlayson, B.L.; McMahon, T.A. Updated world map of the Köppen-Geiger climate classification. *Hydrol. Earth Syst. Sci.* **2007**, *11*, 1633–1644. [CrossRef]
41. Chmielewski, F.M.; Rötzer, T. Response of tree phenology to climate change across Europe. *Agric. For. Meteorol.* **2001**, *108*, 101–112. [CrossRef]
42. Moser, A.; Uhl, E.; Rötzer, T.; Biber, P.; Dahlhausen, J.; Lefer, B.; Pretzsch, H. Effects of Climate and the Urban Heat Island Effect on Urban Tree Growth in Houston. *Open J. For.* **2017**, *7*, 428–445. [CrossRef]
43. De Martonne, E. Une novelle fonction climatologique: L'indice d'aridité. *La Météorol.* **1926**, *21*, 449–458.
44. Chittendon, F. *RHS Dictionary of Plants Plus Supplement*; Oxford University Press: Oxford, UK, 1951–1956.
45. Matsuda, K.; Shibuya, M.; Koike, T. Maintenance and rehabilitation of the mixed conifer-broadleaf forests in Hokkaido, northern Japan. *Eur. J. For. Res.* **2002**, *5*, 119–130.
46. Sugai, T.; Kitao, M.; Watanabe, T.; Koike, T. Can needle nitrogen content explain the interspecific difference in ozone sensitivities of photosynthesis between Japanese larch (*Larix kaempferi*) and Sakhalin fir (*Abies sachalinensis*)? *Photosynthetica* **2019**, *57*, 540–547. [CrossRef]

47. Yamaguchi, M.; Watanabe, M.; Matsumura, H.; Kohno, Y.; Izuta, T. Experimental studies on the effects of ozone on growth and photosynthetic activity of Japanese forest tree species. *Asian J. Atmos. Environ.* **2011**, *5*, 65–67. [CrossRef]
48. Zhang, D.; Katsuki, T.; Rushforth, K. Abies sachalinensis. *IUCN* **2013**. [CrossRef]
49. Cook, E.R.; Briffa, K.R.; Meko, D.M.; Graybill, D.A.; Funkhouser, G. The "segment length curse" in long tree-ring chronology development for palaeoclimatic studies. *Holocene* **1995**, *5*, 229–237. [CrossRef]
50. Fritts, H.C. *Tree Rings and Climate*; Academic Press: London, UK; New York, NY, USA; San Francisco, CA, USA, 1976.
51. Schweingruber, F.H. *Tree Rings. Basics and Applications of Dendrochronology*; D. Reidel Publishing Company: Dordrecht, The Netherlands, 1988.
52. Bunn, A. Dendrochronology Program Library in R. *Dendrochronologia* **2015**, *26*, 115–124. [CrossRef]
53. R Core Team. *R: A Language and Environment for Statistical Computing*; R Foundation for Statistical Computing: Vienna, Austria, 2019.
54. Nowak, D.J.; Kuroda, M.; Crane, D.E. Tree mortality rates and tree population projections in Baltimore, Maryland, USA. *Urban For. Urban Green.* **2004**, *2*, 139–147. [CrossRef]
55. Skiera, B.; Moll, G. The sad state of city trees. *Am. For. (March/April)* **1992**, *98*, 61–64.
56. Hiura, T.; Sano, J.; Konno, Y. Age structure and response to fine-scale disturbances of Abies sachalinensis, Picea jezoensis, Picea glehnii, and Betula ermanii growing under the influence of a dwar bamboo understory in northern Japan. *Can. J. For. Res.* **1996**, *26*, 289–297. [CrossRef]
57. Pretzsch, H. *Modellierung des Waldwachstums*; Parey Buchverlag, Blackwell Wissenschafts-Verlag GmbH: Berlin, Germany, 2001.
58. Hasenauer, H. Dimensional relationship of open-grown trees in Austria. *For. Ecol. Manag.* **1997**, *96*, 197–206. [CrossRef]
59. Umeki, K. Growth characteristics of six tree species on Hokkaido Island, northern Japan. *Ecol. Res.* **2001**, *16*, 435–450. [CrossRef]
60. Peper, P.J.; McPherson, E.G.; Mori, S.M. Predictive equations for dimensions and leaf area of coastal Southern California street trees. *J. Arboric.* **2001**, *27*, 169–180.
61. Schweingruber, F.H. *Tree Rings and Environment*; Swiss Federal Institute for Forest, Snow and Landscape Research: Birmensdorf, Switzerland, 1996.
62. Funakoshi, S. Shoot growth and winter bud formation in *Abies sachalinensis* MAST. *Res. Bull. Coll. Exp. For. Hokkaido Univ.* **1985**, *42*, 785–808.
63. Moser, A.; Uhl, E.; Rötzer, T.; Biber, P.; Caldentey, J.M.; Pretzsch, H. Effects of climate trends and drought events on urban tree growth in Santiago de Chile. *Cienc. Investig. Agrar.* **2018**, *45*, 35–50. [CrossRef]
64. IPCC. *Synthesis Report*; Cambridge University Press: Cambridge, UK, 2014.
65. Rötzer, T.; Rahman, F.; Moser, A.; Pauleit, S.; Pretzsch, H. Process based simulation of tree growth and ecosystem services of urban trees under present and future climate conditions. *Sci. Total Environ.* **2019**, *676*, 651–664. [CrossRef] [PubMed]
66. Temperate Plants Database. Ken Fern. Temperate.Theferns.info. Available online: http://temperate.theferns.info/plant/Taraxacum+officinale (accessed on 8 April 2019).
67. Aihara, K.; Aso, T.; Takeda, M.; Koshiji, T. Actual condition of forest decline and approach (II) The phenomena of forest decline at Tanzawa Mountain in Kanagawa (in Japanese) prefecture. *J. Jpn. Soc. Atmos. Environ. Plan. A* **2004**, *39*, A29–A39.
68. Kohno, Y.; Matsumura, H.; Ishii, T.; Izuta, T. Establishing critical levels of air pollutants for protecting East Asian vegetation—A challenge. In *Plant Responses to Air Pollution and Global Change*; Omasa, K., Nouchi, I., De Kok, L.J., Eds.; Springer: Tokyo, Japan, 2005; pp. 243–250.
69. Kume, A.; Numata, S.; Watanabe, K.; Honoki, H.; Nakajima, H.; Ishida, M. Influence of air pollution on the mountain forests along the Tateyama-Kurobe Alpine route. *Ecol. Res.* **2009**, *24*, 821–830. [CrossRef]
70. Takeda, M.; Aihara, K. Effects of ambient ozone concentrations on beech (*Fagus crenata*) seedlings in the Tanzawa Mountains, Kanagawa Prefecture, Japan (in Japanese with English summary). *J. Jpn. Soc. Atmos. Environ.* **2007**, *42*, 1007–1117.
71. Furukawa, A. Inhibition of photosynthesis of *Populus euramericana* and *Helianthus annuus* by SO_2, NO_2 and O_3. *Ecol. Res.* **1991**, *6*, 79–86. [CrossRef]

72. Yonekura, T.; Yoshidome, M.; Watanabe, M.; Honda, Y.; Ogiwara, I.; Izuta, T. Carry-over effects of ozone and water stress on leaf phenological characteristics and bud frost hardiness of *Fagus crenata* seedlings. *Trees* **2004**, *18*, 581–588. [CrossRef]
73. Gregg, J.W.; Jones, C.G.; Dawson, T.E. Urbanization effects on tree growth in the vicinity of New York City. *Nature* **2003**, *424*, 183–187. [CrossRef] [PubMed]

© 2019 by the authors. Licensee MDPI, Basel, Switzerland. This article is an open access article distributed under the terms and conditions of the Creative Commons Attribution (CC BY) license (http://creativecommons.org/licenses/by/4.0/).

Article

Modeling Ecosystem Services for Park Trees: Sensitivity of i-Tree Eco Simulations to Light Exposure and Tree Species Classification

Rocco Pace [1], Peter Biber [2], Hans Pretzsch [2] and Rüdiger Grote [1,*]

[1] Karlsruhe Institute of Technology (KIT), Institute of Meteorology and Climate Research, Atmospheric Environmental Research (IMK-IFU), Kreuzeckbahnstraße 19, 82467 Garmisch-Partenkirchen, Germany; rocco.pace@kit.edu
[2] Chair for Forest Growth and Yield Science, Faculty of Forest Science and Resource Management, Technical University of Munich, Hans-Carl-von-Carlowitz-Platz 2, 85354 Freising, Germany; peter.biber@lrz.tu-muenchen.de (P.B.); hans.pretzsch@lrz.tu-muenchen.de (H.P.)
* Correspondence: ruediger.grote@kit.edu

Received: 18 January 2018; Accepted: 10 February 2018; Published: 13 February 2018

Abstract: Ecosystem modeling can help decision making regarding planting of urban trees for climate change mitigation and air pollution reduction. Algorithms and models that link the properties of plant functional types, species groups, or single species to their impact on specific ecosystem services have been developed. However, these models require a considerable effort for initialization that is inherently related to uncertainties originating from the high diversity of plant species in urban areas. We therefore suggest a new automated method to be used with the i-Tree Eco model to derive light competition for individual trees and investigate the importance of this property. Since competition depends also on the species, which is difficult to determine from increasingly used remote sensing methodologies, we also investigate the impact of uncertain tree species classification on the ecosystem services by comparing a species-specific inventory determined by field observation with a genus-specific categorization and a model initialization for the dominant deciduous and evergreen species only. Our results show how the simulation of competition affects the determination of carbon sequestration, leaf area, and related ecosystem services and that the proposed method provides a tool for improving estimations. Misclassifications of tree species can lead to large deviations in estimates of ecosystem impacts, particularly concerning biogenic volatile compound emissions. In our test case, monoterpene emissions almost doubled and isoprene emissions decreased to less than 10% when species were estimated to belong only to either two groups instead of being determined by species or genus. It is discussed that this uncertainty of emission estimates propagates further uncertainty in the estimation of potential ozone formation. Overall, we show the importance of using an individual light competition approach and explicitly parameterizing all ecosystem functions at the species-specific level.

Keywords: air pollution removal; BVOC emission; carbon sequestration; tree competition; urban forest

1. Introduction

Population growth, climate change, and high and increasing air pollution levels are known to pose risks to health and safety in cities [1]. Therefore, sustainable urban planning is necessary to improve the quality of life and preserve the integrity of natural ecosystems [2]. Urban forests and trees can significantly contribute to mitigating climate change effects and improving air quality in residential areas [3]. Pollution caused by tree removal in cities considerably depends on the species and their properties [4]. For this reason, species selection is important to achieve optimal results of city greening [5]. To find the most suitable trees or tree mixture, ecosystem services models can be particularly useful as decision-support tools for city planning [6,7].

The i-Tree Eco model is an ecosystem service model for urban trees developed by the US Department of Agriculture (USDA) Forest Service for application in the U.S., and it has been adopted by the U.K., Australia, and Canada [8,9]. The model is widely used to evaluate urban vegetation-induced environmental services [10–12], e.g., carbon storage and sequestration, air pollution reduction, and water runoff reduction, the effects of trees on energy consumed by buildings, and some disservices, such as the emission of biogenic volatile organic compounds (BVOCs).

The i-Tree Eco requires information concerning the species and the stem diameter at breast height (DBH) as the input data. Additional data, including land use criteria, total tree height, crown size (height to live top, height to the crown base, crown width, and percentage of crown missing), crown health (dieback or condition), and competition status, can improve the model accuracy. Most of these input data are usually determined in the field by explicit visual inventories. This determination method is relatively easy to learn but remains subjective and prone to errors. For large areas, sample plots are required to be investigated and scaled to the whole region, leading to considerable uncertainties when the species distribution is non-homogeneous [13]. In addition, the efforts involved in defining the sample plots, educating the field researchers, and applying the inventory requires more time and money [14] compared to the case wherein the i-Tree Eco protocol, which refers to the data available from remote sensing or GIS, is used. The latter facilitates the assessment of pollution reduction in cities using the input of leaf area index (LAI) as a structural characteristic [15–18]. However, it is worth noting that in many cases, urban forest inventories are unsuitable for deriving all the required parameters, e.g., crown light exposure (CLE), making the use of default model parameterization for many tree properties more appealing.

Species characterization data is important for i-Tree Eco initialization because these data define the basic parameters used for calculating all ecosystem services and disservices (e.g., leaf area (LA), leaf biomass, allometric equations, BVOCs emission rates). If the species information is not available, the i-Tree Eco model uses values that are defined by the genus, family, or type (evergreen/deciduous). This may be particularly problematic if the model is applied to regions other than the U.S., for which the parameters were designed. Another important characterization is the degree of competition that a tree experiences. This is expressed as "CLE" in the i-Tree Eco model based on a visual estimate according to Bechtold (2003) [19]. Because this parameter is evaluated by educated investigators, it is expensive and remains subjective. The default value for the competition is differentiated into street trees and trees within urban parks and forests but is independent of size or closeness to neighbors. Hence, it is desirable to define i-Tree Eco initialization for species, size, and composition of each tree using cost-effective and objective methods, particularly when large areas are concerned.

A promising approach to achieve an automated initialization is the application of remote sensing methods, which has been attempted in recent investigations [15,20–22]. However, because of the complexity associated with urban areas (e.g., high spectral similarity of vegetation types or overlooked small trees in high-density stands), initializations using remote sensing data have an inevitable degree of uncertainty concerning species differentiation [23]. Additionally, it should be considered that most urban trees are deciduous [4,24]. This creates difficulties in species distinction because the photogrammetric interpretation of aerial photographs generally allows only the separation of evergreen from deciduous trees, particularly when the plant species diversity is as high as that in the urban context [25]. Furthermore, although tree position and size can be reasonably well determined, automatic information on light competition between trees is not provided by these approaches [26,27].

The actual competition between trees has been estimated using multiple approaches. These estimations include the influence zone of each tree and the degree and nature of interaction [28]; a competition algorithm based on light intensity [29]; the "moving average autoregression" method to assess the spatial dependence attributable to the competition and micro-site influences [30]; and the calculation of the inter-tree competition between each tree based on the position, height, and crown size of a tree and its competitors [31]. The aforementioned methods provide objective measures because they are based on tree position and size, i.e., the remote sensing data.

Herein, we propose a methodology to derive competition data from dimensional tree information and investigate uncertainties inherent in i-Tree Eco model applications related to species determination. Therefore, we apply the model to Englischer Garten—a large urban park in Munich, Germany—where detailed inventory data are available, to assess how the estimates of ecosystem services and disservices change when we gradually decrease the degree of initialization detail. In particular, we hypothesized that accurate species information is crucial for determining several ecosystem processes and functions.

2. Materials and Methods

2.1. Site Description and Model Inputs

In this study, we analyzed the south of "Englischer Garten", a 330-ha park located in Munich, Germany (Figure 1). The site is mainly comprised of a mix of deciduous trees dominated (≈77%) by the species Norway maple (*Acer platanoides* L.), European beech (*Fagus sylvatica* L.), small-leaved lime (*Tilia cordata* MILL.), sycamore maple (*Acer pseudoplatanus* L.), and European ash (*Fraxinus excelsior* L.). Evergreen species contribute less than 1% to the total tree number (Table 1). The inventory comprises 9391 trees growing within the "Englischer Garten" park, which have been collected by the Bavarian Administration of State-Owned Palaces, Gardens and Lakes. These tree data (species, tree height, DBH, crown diameter, and height to the crown base) were used as input parameters in i-Tree Eco to calculate the ecosystem services provided by the park. Since information about the crown condition has not been collected, we assumed a crown condition involving the best tress, i.e., 0% crown missing and 100% healthy (or 0% dieback). Further, these input parameters can be used for increasing the accuracy of LA and carbon sequestration estimations [32].

Table 1. Species composition of the south part of Englischer Garten (a total of 9391 trees).

Species	Relative Number (%)	Basal Area (m^2)
Norway maple (*Acer platanoides* L.)	32.9	453.6
European beech (*Fagus sylvatica* L.)	12.6	447.4
Small-leaved lime (*Tilia cordata* MILL.)	11.2	160.1
Sycamore maple (*Acer pseudoplatanus* L.)	10.4	122.8
European ash (*Fraxinus excelsior* L.)	10.1	223.2
Field maple (*Acer campestre* L.)	3.7	35.8
European hornbeam (*Carpinus betulus* L.)	3.6	43.6
Horse-chestnut (*Aesculus hippocastanum* L.)	2.6	70.1
English oak (*Quercus robur* L.)	2	32.4
Scotch elm (*Ulmus glabra* Huds.)	1.8	28.8
Black locust (*Robinia pseudoacacia* L.)	1.6	21.7
London plane (*Platanus* × *acerifolia* Aiton)	1.3	19.4
White willow (*Salix alba* L.)	1.0	43.3
Willows (*Salix* spp.), poplars (*Populus* spp.), cherries (*Prunus* spp.), Caucasian wingnut (*Pterocarya fraxinifolia*), birches (*Betula* spp.), hazels (*Corylus* spp.), walnuts (*Juglans* spp.), common pear (*Pyrus communis*), honey locust *Gleditsia triacanthos*, tulip tree (*Liriodendron tulipifera*), hawthorns (*Crataegus* spp.), ginkgo (*Ginkgo biloba*), whitebeams (*Sorbus* spp.), grey alder (*Alnus incana*), tree of heaven (*Ailanthus altissima*), cornelian cherry (*Cornus mas*), Japanese pagoda tree (*Sophora japonica*), yew (*Taxus baccata*), pines (*Pinus* spp.), and spruce (*Picea abies*), magnolia (*Magnolia* spp.)	5.2 (evergreen species <1%)	82.6

The model simulations were conducted in 2012 using hourly meteorological data registered at the Munich Airport (temperature, wind speed, and radiation from the National Oceanic and Atmospheric Administration (NOAA, Silver Spring, MD, USA) database that is directly accessed by i-Tree Eco, Farnham, UK) and the München Theresienstrasse (precipitation) weather stations, which are about 40 and 1 km away from the study site, respectively. The average hourly concentration data for ozone (O_3), sulfur dioxide (SO_2), nitrogen dioxide (NO_2), carbon monoxide (CO), and particulate matter with a diameter of 2.5 ($PM_{2.5}$) were provided from the Bavarian Environment Agency (LFU), Augsburg,

Germany. PM_{10} data are no longer analyzed in the latest i-Tree Eco version (version 6) because $PM_{2.5}$ is generally more relevant for human health [33]. This information was checked for quality and fed into the USDA Forest Service database for its use in our investigation.

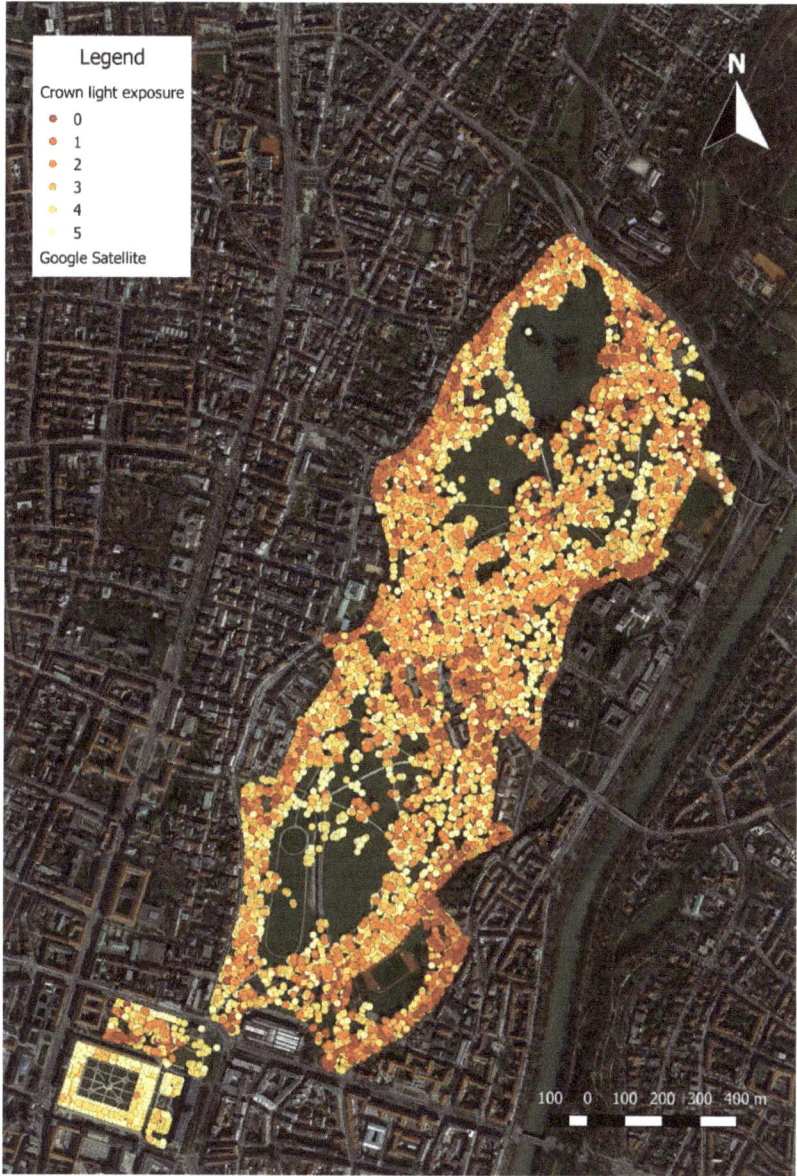

Figure 1. All trees in the south of Englischer Garten with relative crown light exposure (CLE) classes calculated by the CCS (crown competition for sunlight) competition index which is based on the algorithms provided in Pretzsch et al. [31]. In addition to competition between trees, shading by buildings has been considered as well.

2.2. CLE Effects

The i-Tree Eco model considers the average competition of a tree as the degree of crown exposure to sunlight, which is not supposed to change during the simulation. This competition is expressed as CLE (crown light exposure), which is an empirical index that reflects the number of sides of a tree receiving direct sunlight [32]. Therefore, the tree crown is virtually divided into the four cardinal directions and an additional surface on top of the crown (Figure 2) [19]. A classification can thus result in a CLE value between 0 (which would characterize a fully suppressed tree in the understorey of a closed canopy, only receiving diffuse light) and 5 (solitary tree not shaded by surrounding trees or other obstacles). CLE reflects broadly the capability for photosynthesis and is used to calculate LA and tree growth estimates. While growth directly determines carbon sequestration, LA influences various ecosystem services. Herein, we investigate air pollution reduction and biogenic emissions to determine the CLE sensitivity (see Section 2.4).

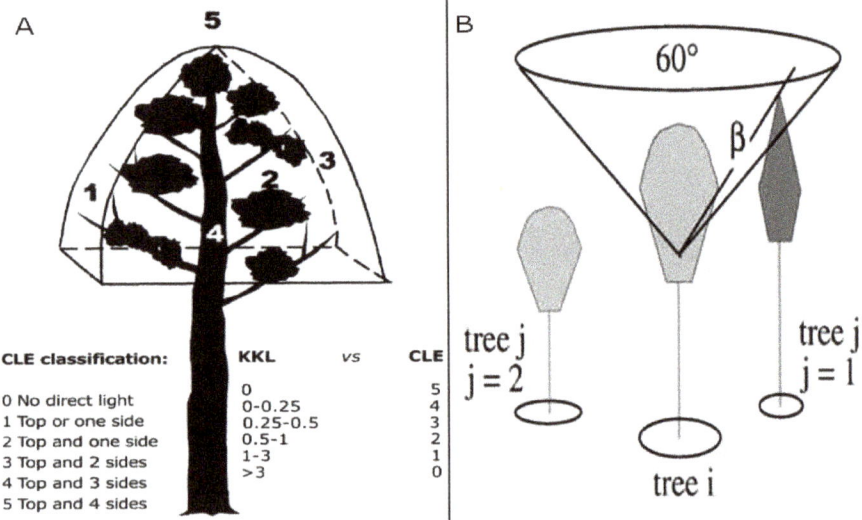

Figure 2. Crown light exposure (CLE) classification (**A**) and calculation of the competition index CCS (crown competition for sunlight) based on Pretzsch et al. (**B**). In the bottom right corner of (**A**), the conversion of CCS is indicated in CLE values. Sources: [19,31].

2.2.1. Calculation of Leaf Area (LA)

The i-Tree Eco model uses tree-specific CLE values that are grouped into three states in order to calculate LA: the open-grown (CLE = 4–5), park (CLE = 2–3), and closed forest (CLE = 0–1) conditions. Under the open-grown condition, LA is calculated either from DBH only (measured at 1.37 m above ground) or from crown length (H) and crown width (D) if available [34]:

- CLE = 4–5 (open-grown trees)

$$\ln(LA) = b_0 + b_1 DBH + b_2 S \quad (1)$$

$$\ln(LA) = b_0 + b_1 H + b_2 D + b_3 S + b_4 C \quad (2)$$

In these calculations, S is a species-specific shading factor, which is defined as the percentage of light intensity intercepted by foliated tree crowns, and C is the outer surface area of the tree crown calculated from H and D as $C = \pi D(H + D)/2$. S varies with species for deciduous trees, and if it is not defined for individual species, the averages for the genus or general hardwoods are used. For conifer trees, the model applies a shading factor of 0.91 for all species, except for pines (0.83) [32,34]. For the closed forest condition, LA is calculated using the following equation based on the Beer-Lambert law:

- CLE = 0–1 (forest stand condition)

$$LA = (\ln(1 - S)/-k) \times \pi \times (D/2)^2 \quad (3)$$

where k is a light extinction coefficient that is differentiated between conifers (0.52) and hardwoods (0.65) [32]. For CLE = 2–3 (park condition), LA is calculated as the average value determined by the open-grown (CLE = 4–5) and closed canopy equations (CLE = 0–1).

2.2.2. Effects on Tree Growth

Average diameter growth is added to tree diameter (year x) to estimate the tree diameter in year x + 1 [35]. In i-Tree Eco, a standard diameter growth (SG) that can be reduced is defined for open-grown trees (CLE = 4–5) when the number of frost-free days is smaller than a defined value:

$$\text{Standard diameter growth (SG)} = 0.83 \text{ cm/year} \times (\text{number of frost-free days}/153) \quad (4)$$

Park tree growth (CLE = 2–3) is calculated by dividing SG of open-grown trees by 1.78 and that of forest trees (CLE = 0–1) by 2.29.

2.2.3. Automated Competition Calculations

Herein, we calculated CLE based on a routine introduced within the framework of the single-tree-based stand simulator SILVA [31]. This model calculates single-tree growth in relation to its surrounding three-dimensional space to produce the competition index CCS value. CCS aids in identifying the competitors of single trees by considering a virtual reverse cone with an axis equal to the tree axis and its vertex placed within the crown of the tree (Figure 2b). The model determines the angle β between the insertion point of the cone and the top of any competitor tree. This angle is multiplied by the crown cross-sectional areas (CCAs) of the competitors and the tree of interest considering a species-specific light transmission coefficient:

$$CCS_i = \sum_{j=1}^{n} \left(\beta_i \frac{CCA_j}{CCA_i} TM_j \right) \quad (5)$$

where CCS_i is the competition index for tree i; β_i is the angle between cone vertex and top of competitor j; CCA_j and CCA_i are the CCAs of trees j and i, respectively; TM_j is the species-specific light transmission coefficient for tree j; n is the number of competitors of tree i [31].

We calculated the competition index CCS for each tree before converting the results into a CLE classification. Therefore, we assumed that the trees without competition (CCS = 0) correspond to the highest CLE value (CLE = 5), while other CLE classes are assigned to CCS values according to the relative abundance of trees (see Figure 2a). Thus, there is approximately the same number of trees in each CLE class. We also considered shading by buildings and other trees not included in our inventory using the open source Quantum GIS (QGIS) software and a fixed buffer distance of 15 m around the trees. For all trees falling within this distance, we reduced the CLE by one unit.

2.3. Model Sensitivity Studies

The importance of individually determined CLEs was tested by comparing simulated carbon sequestration, pollution reduction, and emissions calculated using the SILVA routine (indicated as individual CLE runs) with the results originating from default CLE values (indicated as average CLE runs). The sensitivity connected to uncertain tree species determination was also investigated by deriving parameters from different information sources: (1) individually determined tree species through field measurements (species-specific runs); (2) assuming a genus-specific inventory that reflects the detail derived by, for example, sophisticated remote sensing techniques, such as LiDAR (light detection and ranging) and high-resolution images (genus-specific runs) [23]; and (3) assuming a distinction between evergreen and deciduous trees representing the detail that can be derived by photogrammetric interpretation of aerial photographs (dominant species runs) [25]. In this context, we hypothesized that all deciduous species were *A. platanoides*, the dominant species of the park (Table 1), and all evergreen trees had the properties of *Picea abies*, the most common conifer species in South Germany [36].

2.4. Model Calculations

For all calculations, the initial dimensional data, hourly temperature and precipitation records, and air pollution concentrations are used as input data [32,37]. Only ecosystem services with a direct link to air chemistry, i.e., air pollution reduction, carbon sequestration, and biogenic emissions, were evaluated. In subsequent paragraphs, we describe how the model calculates these services.

2.4.1. Carbon Storage and Sequestration

The model calculates the above-ground biomass of trees in dry weight using allometric equations [35,38] and the total tree biomass using a root-to-shoot ratio of 0.26 [32,39]. Because the allometric equations are diameter-based and developed for closed canopies, biomass should be corrected for open-grown trees, which tend to be shorter and thus have less above-ground biomass at a given diameter. Therefore, biomass estimates are reduced for urban trees by a factor of 0.8 [38]. Total carbon storage is calculated by multiplying tree dry weight biomass by 0.5 [32]. Hence, assuming that there is no change in soil carbon, annual carbon sequestration is directly calculated from tree growth (see Section 2.2.2) [32].

2.4.2. Air Pollution Reduction

The dry deposition of O_3, SO_2, NO_2, CO, and $PM_{2.5}$ is hourly determined throughout the year [40–43]. During precipitation events, deposition is assumed to be zero. Using the following equation, for other periods, the pollutant flux into the biosphere (F; in g m^{-2} s^{-1}) is calculated as the product of deposition velocity (V_d; in m s^{-1}) and pollutant concentration (C; in g m^{-3}) [41]:

$$F = V_d C \qquad (6)$$

Using the following equation, the deposition velocities of CO, NO_2, SO_2, and O_3 are calculated as the inverse of the sum of the aerodynamic resistance R_a, a quasi-laminar boundary layer (R_b), and the canopy resistance R_c expressed in s m^{-1} [44]:

$$V_d = (R_a + R_b + R_c)^{-1} \qquad (7)$$

where R_a is determined from meteorological data (wind speed and atmospheric stability) since it is assumed to be independent of air pollution type or plant species, R_b is based on a value defined in a study by Pederson et al. (1995) [45] using a specific Schmidt number for each air pollutant [46], and R_c is calculated using the following equation:

$$1/-R_c = 1/(r_s + r_m) + 1/r_{soil} + 1/r_t \qquad (8)$$

where r_s is the stomatal resistance (s m^{-1}), r_m is the mesophyll resistance (s m^{-1}), r_{soil} is the soil resistance (2941 s m^{-1} in growing season and 2000 otherwise), and r_t is the cuticular resistance (s m^{-1}).

Hourly canopy resistance values for O_3, SO_2, and NO_2 were calculated based on a modified hybrid of big leaf and multilayer canopy deposition models [44,47]. The model calculates stomatal resistance (r_s) as the inverse of stomatal conductance, which is estimated based on the leaf photosynthetic rate, relative humidity, and surface CO_2 concentration using the Ball-Berry formula (for more details, see [46]). The mesophyll and cuticular resistance values are set based on those reported in the literature: for NO_2, r_m = 100 s m^{-1} [48] and r_t = 20,000 s m^{-1} [49]; for O_3, r_m = 10 s m^{-1} [48] and r_t = 10,000 s m^{-1} [50,51]; and for SO_2, r_m = 0 [49] and r_t = 8000 s m^{-1}. As CO reduction is assumed to be independent of photosynthesis and transpiration, the resistance value for CO is set to 50,000 s m^{-1} in the in-leaf season and 1,000,000 s m^{-1} in the out-leaf season for all trees [52]. The hourly deposition and resuspension rates for $PM_{2.5}$ are calculated based on wind speed and LA (for more details, see [42]).

The base deposition velocity V_d was multiplied by the individual tree LAI based on local field data and local seasonal variation (local leaf-on and leaf-off dates). For deciduous trees, the calculation of pollution deposition is limited to the in-leaf period. Herein, the leaf-on date is 5 April, whereas the leaf-off date is October 28.

2.4.3. Biogenic Emissions

Hourly emissions of isoprene (C_5H_8) and monoterpenes (C_{10} terpenoids) were estimated using an approach proposed by Guenther et al. 1993 [53] and Geron et al. 1994 [54] with genus-specific parameters [55]. In this approach, emission was calculated by multiplying leaf biomass (derived from LA with species-specific conversion factors) by emission rates. These in turn depend on temperature and light (isoprene) or temperature only (monoterpenes) as well as on genus-specific factors that represent emissions at 30 °C and 1000 µmol m^{-2} s^{-1} photosynthetically active radiation (PAR) [56]. Median emissions values for the family, order, or superorder were used if genus-specific emission was not available [32]. Incoming PAR was calculated as 46% of total solar radiation input [57]. Because isoprene emission has a nonlinear dependence on light, PAR was estimated from incoming PAR for 30 canopy levels using the sunfleck canopy environment model with the LAI of the analyzed structure [32]. Hourly leaf temperature was calculated from air temperature while considering the transpiration rate (for unlimited water supply), LAI, and percentage tree cover [55].

3. Results

3.1. Ecosystem Services

The tree crowns of Englischer Garten were calculated to cover an area of 73.2 ha and have a total LA of 467.5 ha with a leaf biomass of 301.7 tons. The most dominant species in terms of number and basal area were *A. platanoides* and *F. sylvatica* (Table 1). Overall, carbon stored in the trees was estimated to be 6225 tons. In accordance with their fraction of basal area, *F. sylvatica* and *A. platanoides* stored the most carbon (31.4% and 29.1% of the total, respectively). The amount of carbon sequestered in 2012 was calculated to be 214 tons (Figure 3). The model further indicates that the trees in Englischer Garten removed 2610 kg of O_3, 845 kg of NO_2, 186 kg of $PM_{2.5}$, 171 kg of SO_2, and 62 kg of CO (Figure 4). In addition, the trees emitted an estimated BVOC amount of 550 kg (158 kg isoprene and 392 kg monoterpenes; Figure 5).

3.2. Sensitivity to Average CLE Values

Simulation with average CLE values (CLE average) resulted in a 14% reduction in LA and leaf biomass (379.9 vs. 443 ha year^{-1}; 224.9 vs. 262 t year^{-1}) compared to the run with individual determination of CLE (Figure 3). This directly affected bioemissions (494 vs. 550 kg year^{-1}; Figure 5) and pollution reduction (Figure 4). Except for CO (in both cases, 62 kg year^{-1}), more pollutants were removed in the "CLE individual" simulation compared to the "CLE average" simulation (2610 vs. 2482 kg year^{-1} for O_3, 845 vs. 794 kg year^{-1} for NO_2, 186 vs. 165 kg year^{-1} for $PM_{2.5}$, and 171 vs. 164 kg year^{-1} for SO_2). Carbon sequestration was also affected by individual light exposure (Figure 3): 214 tons of carbon (784 CO_2 equivalent) removed in the "CLE individual" simulation compared to 155 tons of carbon (567 CO_2 equivalent) removed in the "CLE average" simulation, indicating that the trees in Englischer Garten were more open grown than expressed by the average CLE value.

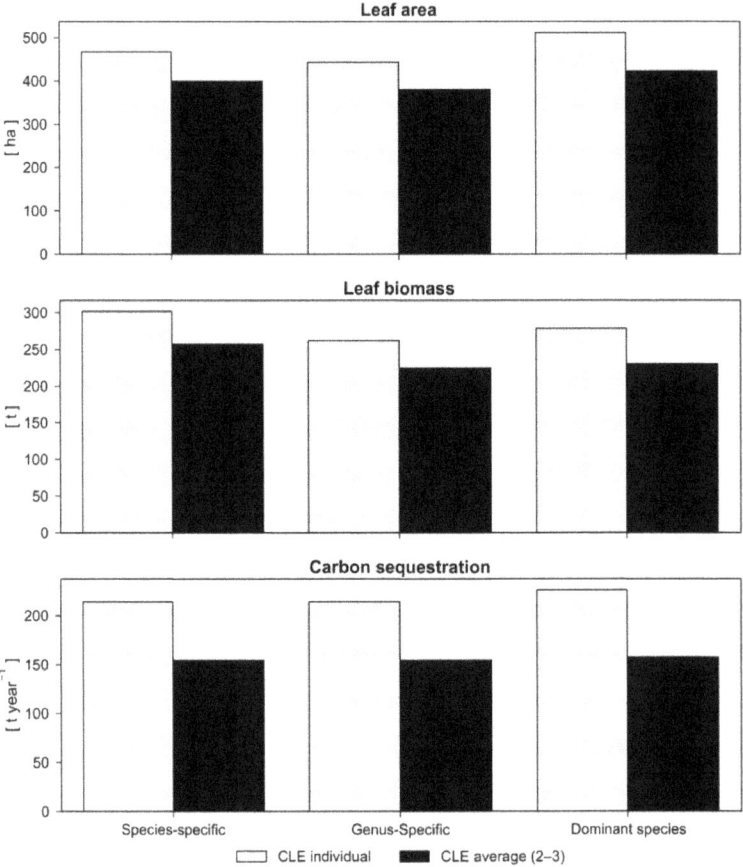

Figure 3. Leaf area (LA), leaf biomass, and carbon sequestration comparison between the "species-specific", "genus-specific", and "dominant species" simulations considering the crown light exposure (CLE) value for each tree (CLE individual) and the average CLE values (CLE average).

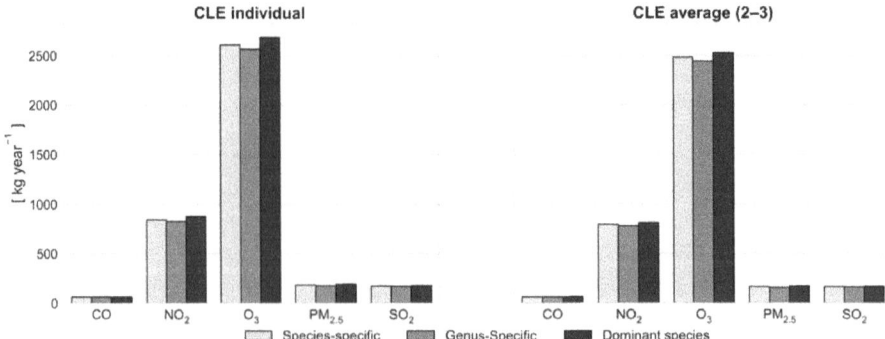

Figure 4. Air pollution reduction comparison between the "species-specific", "genus-specific", and "dominant species" simulations considering the crown light exposure (CLE) for each tree (CLE individual) and the average CLE values (CLE average).

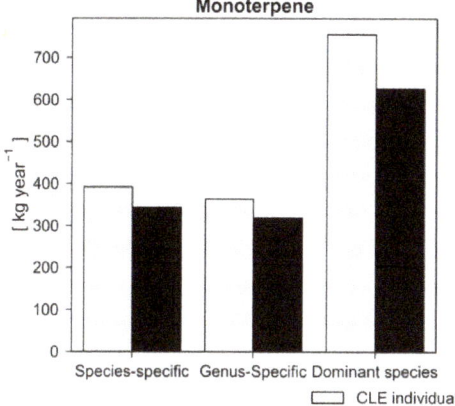 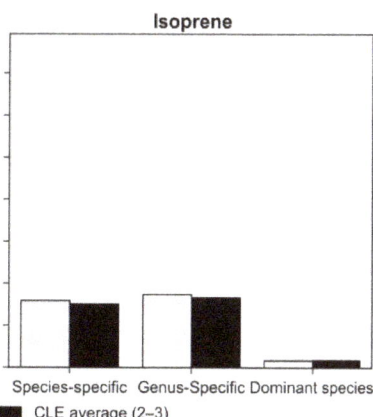

Figure 5. Comparison of monoterpene and isoprene emissions between the "species-specific", "genus-specific", and "dominant species" simulations considering the crown light exposure (CLE) for each tree (CLE individual) and the average CLE values (CLE average).

3.3. Sensitivity to Different Species Classification

Compared to simulations with species-specific parameterization, the results obtained with the genus-specific simulation showed lower LA (−5%; 443.7 ha) and leaf biomass (−13%; 262 t). The dominant species simulation demonstrated higher LA values (511.3 ha; 9% higher than the species-specific parameterization), but leaf biomass results (−8%; 278.4 t) were almost equivalent (Figure 3).

According to lower LA, pollutants removed with genus-specific parameterization were fewer than those removed with species-specific parameterization (−2% of O_3, −2% of NO_2, −5% of $PM_{2.5}$, and −1% of SO_2), except for CO (62 kg year^{-1}), as shown in Figure 4. The effects of species differentiation were small. LA also had a minor effect on BVOC emissions, which are otherwise driven by marginally different parameters for the species- and genus-specific simulations. However, when the parameters were set according to the dominant species approach, monoterpene emissions were considerably higher (+93%) and isoprene emissions tended to be zero (−89%) compared to the species-specific simulation.

Carbon sequestration (214 t) was not altered by the species- and genus-specific parameterizations because biomass growth was found to solely depend on the tree size and competition state (Figure 3). Instead, the dominant species simulation showed higher carbon sequestration (+5.7%; 226 t). In this case, a larger number of trees fell into class 4–5 due to the smaller transmission coefficients for maples in comparison with the species average (Figure 6).

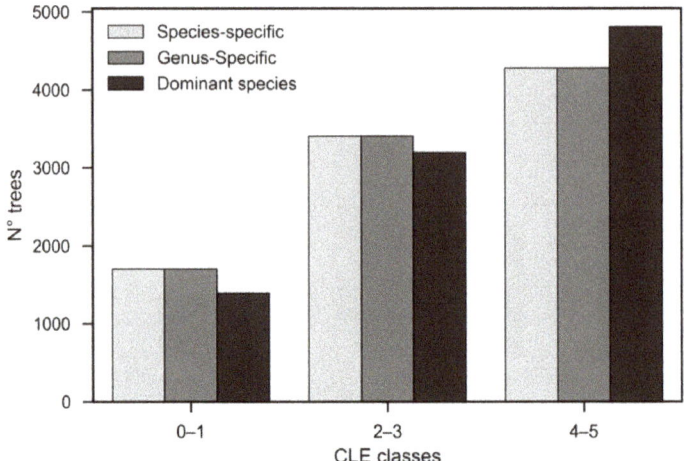

Figure 6. Comparison of crown light exposure (CLE) classes between the "species-specific", "genus-specific", and "dominant species" simulations. Class 4–5 (open-grown trees) is dominant in all simulations, particularly in the "dominant-species" simulation (99.4% of *Acer platanoides*).

4. Discussion

Our investigation demonstrates that Englischer Garten—a major park in Munich—provides a considerable amount of environmental services, i.e., air pollution reduction and carbon sequestration. Simulated removals of O_3 and NO_2 were at the highest rates (3.6 and 1.1 g m^{-2}, respectively). A comparison with the estimates for other European cities showed that the total pollutant (combining CO, O_3, NO_2, SO_2, and $PM_{2.5}$) removal rate, which was 5.3 g m^{-2} of tree cover per year, was similar to that for the city of Strasbourg (5.1 g m^{-2}) [12] but lower than that for London (8.7 g m^{-2}) [58]. The simulated $PM_{2.5}$ removal rate for the park (0.25 g m^{-2}) was of the same magnitude as that determined for U.S. cities (0.25 g m^{-2}) [42], but the total removal rate (considering $PM_{2.5}$ instead of PM_{10}) was lower than the average value calculated for the U.S. cities (7.5 g m^{-2}) [41].

Although we focused on uncertainties associated with parameterization and the determination of light competition (discussed further below), we also addressed the process uncertainty that is tightly linked to parameterization because processes and parameters often share the same knowledge base. Other uncertainties associated with the model inputs were climate conditions, which resulted in different degrees of ecosystem services during different years, and air pollution data measured outside the park, which may not reflect the ambient conditions of the park trees. However, since these aspects do not alter the relative contribution to the ecosystem services of different tree species, their investigation was outside the scope of this study.

4.1. Uncertainties Associated with Parameterization

Species definition is fundamental information on which many properties depend. Therefore, we introduced a degree of uncertainty in species definition based on either imperfect knowledge—as is often the case with species in areas outside the original model development region (genus-specific parameterization)—or what can reasonably be derived from remote sensing measures (dominant species parameterization).

The total LA in the "genus-specific" simulation was approximately 5% lower than that in the "species-specific" simulation, which slightly reduced pollution removal. This reflects the similarity between species- and genus-based parameters, exhibiting the almost linear scaling of LA and deposition in i-Tree Eco. The differences are more comprehensively expressed with BVOC emissions that are higher for isoprene (+10%) and lower for monoterpene (−7%) using only genus-specific parameterization. This highlights the additional influence of species-specific conversion factors between LA and leaf biomass. However, there were no differences between the two simulations concerning carbon sequestration because it is based on a fixed diameter growth rate and allometric equations from the literature that are used to calculate biomass change from the dimensional change [38,56]. Such allometric equations are usually derived from forest trees and, although they have been adjusted for urban conditions, they might not exactly reflect the actual tree forms and wood density. For example, various investigations show that high ozone concentration can cause a reduction in the root/shoot ratios of trees [59], especially for deciduous species [60]. Under high air pollution a larger root biomass fraction can thus be expected.

This approach might be improved by introducing a dependency on climate and soil type, particularly by encompassing drought stress events [61] and other stress factors that might depend on the degree of air pollution or salt application. Pretzsch et al. (2017) [62] showed that the urban trees in Europe have accelerated their growth since 1960 because of the effects of climate change, with considerable differences in different climate regions. Dahlhausen et al. (2017) [63] demonstrated the effect of urban climate on lime tree growth in Berlin and found that the trees in the city center were more responsive than those in urban boundaries. It is a challenge to represent these findings with the i-Tree Eco model by only considering the limited number of frost-free days, as is currently the case. Instead, representing the growth demands the introduction of direct sensitivity to temperature, competition, and possibly other factors that will vary from those of the city of interest. In addition, wood density and growth form might require parameterization that is regionally adapted to European species [64,65].

The dominant species parameterization assumes all deciduous trees (99.4% of total population) to be Norway maple (*A. platanoides*) and uses the parameters of spruce (*P. abies*) for all the evergreen trees (0.6%). Although the effects were found to strongly depend on the species composition and the dominant species used for representation, the results showed that tree species misclassification particularly affects BVOC emission estimates. Additionally, important information concerning LA and related air pollution reduction rates was derived. The effect of parameterization on emissions was expressed because emissions factors change strongly between species. For example, high monoterpene emissions resulted from *A. platanoides*—a dominant species having one of the highest monoterpene emission factors (1.6) of all species in the inventory (0.6 for *Fagus*, 0.1 for *Fraxinus*, and 0 for *Tilia*). In contrast, maples had a considerably low isoprene emission factor (0.1) compared to that of a number of species that were neglected in the dominant species runs, reducing the overall emission estimate for this compound. For example, *Quercus robur*, *Robinia pseudoacacia*, *Platanus* × *Acerifolia*, and *Salix alba* were all assumed to be considerably high isoprene emitters (with emission factor equal to 70) [56].

However, it is worth noting that emission parameters are uncertain. For example, monoterpene emission factors reported in the literature vary between 0.1 [66] and 43.5 [67] for *Fagus*, between 0 [68] and 9.6 [69] for *Fraxinus*, and between 0 [70] and 1.2 [71] for *Tilia*. For subtropical street trees, Dunn-Johnston et al. (2016) [72] showed that a considerable difference exists between species-derived isoprene emission rates and those assumed (genus-specific emission rates) in i-Tree Eco. Other uncertainties in the emission pattern of plants were found to originate from seasonal changes that depended on the weather conditions of the previous days and weeks [73], which were not considered herein. Additionally, the potential effects of air pollution [74] or drought [75,76], which might increase or decrease emissions, have been neglected. Since emissions are supposed to trigger aerosol production and ozone formation [77], these deficits may need consideration in order to be appropriately used in combination with regional air chemistry and climate models [78].

4.2. Uncertainties Associated with Competition Calculations

Another form of influential information is the determination of competition, which represents the light that an individual tree is able to utilize during photosynthesis. We demonstrated that neglecting an individual CLE determination (CLE average) strongly affected LA and thus pollution reduction (except for CO), carbon sequestration, and bioemissions. The intensity of this effect depended on the degree to which tree size and position differed from a homogeneous distribution with medium tree distances. Free-standing trees requiring more sunlight are therefore underrepresented in terms of LA and growth when only using average CLE determination [79]. Since Englischer Garten is characterized by many open spaces and areas where trees are found to clump together, the divergence from the mean conditions is relatively large (Figure 6). However, this is not an exception, since parks are often characterized by a complex tree distribution pattern in order to optimize recreational activities [3]. Therefore, we assumed that the structure of the Englischer Garten park was representative and that the significance of considering individual crown exposure was possibly high.

It is worth noting that the uncertainties associated with competition estimates can multiply with those associated with parameter estimation. In our case, the difference between the individual and mean competition calculations was particularly high when only "dominant species" parameters were used: the class with average competition (CLE = 2–3) was 56% larger than that with high competition (CLE = 0–1) and 51% smaller than that with low competition (CLE = 4–5) (50% and 26% for the species- and genus-specific simulations, respectively; Figure 6). Part of this effect was due to the method we designated CLE classes from competition calculations. Because competition depends on transmission properties of the canopy that differ with species, dominant-species parameterization results in different tree numbers per CLE class than species- or genus-specific parameterization (no difference exists between those two since light transmission factors are homogeneous within a genus) [31].

4.3. Uncertainties Associated with Processes

Pollution deposition was calculated from LA and species-specific deposition velocities. However, uptake processes may need to be differentiated in terms of stomatal uptake and deposition onto the surface of the leaves, both of which are related to further environmental conditions. Stomatal uptake was found to depend on the photosynthetic activity, turgor pressure, and removal time in the intercellular spaces, particularly for O_3 and NO_2, which were almost immediately metabolized [4]. Therefore, stomatal uptake through stomata could possibly be influenced by drought stress, which the standard i-Tree Eco model neglects, assuming sufficient water availability or irrigation for the street and park trees. In fact, the stomatal closure has been shown to considerably reduce pollution removal in dry periods and alternative model formulations have been suggested [80,81]. Consequently, surface deposition was found to strongly depend on leaf structure [82,83], which determines both velocity and deposition capacity. Deposition capacity in turn is also dynamical, since washing the deposited content from the leaf reduces the pollutant storage at the leaves [84]. In the case of NO_2 and SO_2, it is also possible that pollution removal occurs by means of dissolving directly into the water film on the plant surface [4]. Although it might have a minor impact compared with the influence of the stomata on the gaseous uptake, the uncertainty associated with leaf-surface properties and the dependency on rainfall are currently underexplored in i-Tree Eco.

In addition, pollutant deposition varies with the canopy structure and tree position because these factors determine the air flow within and around trees [85]. However, the relation between tree traits (e.g., crown geometry and foliage distribution), "urban street canyon", and weather conditions are complex and cannot be evaluated using i-Tree Eco. For example, low turbulence may promote deposition and increase the N_2O and CO concentrations or O_3 formation due to longer residence time [86].

Based on the above discussion, it can be easily observed that i-Tree Eco estimates constitute a relatively first-order approximation that is solely related to surface area and uses lumped velocity parameters for pollutants that are nevertheless species specific. Only CO removal is independent of LA, possibly because it is independent of photosynthesis and transpiration [52,87]. This assumption is likely to be defensible because CO deposition in soil has been demonstrated to be considerably larger than that in plant leaves [48,88].

5. Conclusions

Englischer Garten trees provide important ecosystem services to Munich. The results of our model simulation revealed that in 2012, they potentially removed 4 t of pollutants, particularly ozone and nitrogen dioxide, and more than 200 t of carbon dioxide. The robustness of the model estimates was found to strongly depend on accurate species-specific parametrization as well as an individual determination of the competition state of the trees. The suggested automated determination process represents an objective procedure that is in many ways preferable to manual investigations, although its applicability and precision remains to be tested against field observations. Species-specific parameterization is particularly necessary for the determination of BVOC emissions. Since the taxonomic approach of i-Tree Eco only assigns genus or family values, simulations have uncertainties associated with species that differ from the genus- or dominating species-based settings, even if their abundance is relatively small. Therefore, it is strongly recommended to complement remote sensing applications with terrestrial observations and improve the database of species-specific parameters. Overall, we highlighted a considerable potential for improvements in the i-Tree Eco model process representations, particularly concerning the climate sensitivity of emission and growth processes.

Acknowledgments: This research was supported by Graduate School for Climate and Environment (GRACE). We also acknowledge support by Deutsche Forschungsgemeinschaft and Open Access Publishing Fund of Karlsruhe Institute of Technology. Furthermore, we thank the Bavarian Administration of State-Owned Palaces, Gardens and Lakes for providing the individual tree data, the Bavarian Environment Agency (LFU) for providing air pollution data, the i-Tree support team for assisting us in model implementation and data processing, and the Graduiertenzentrum Weihenstephan of Technical University of Munich and the ENAGO company for the English editing.

Author Contributions: R.P. performed the simulation experiments and analyzed the data; R.P. and R.G. conceived and designed the experiments and jointly wrote the text; P.B. and H.P. provided the competition algorithm, supported the implementation, and revised the text.

Conflicts of Interest: The authors declare no conflict of interest.

References

1. United Nations. *World Urbanization Prospects: The 2014 Revision, Highlights (ST/ESA/SER.A/352)*; Department of Economic and Social Affairs, Population Division: New York, NY, USA, 2014; ISBN 9789211515176.
2. UN-Habitat. *Urbanization and Development: Emerging Futures. World Cities Report 2016*; United Nations Human Settlements Programme: Nairobi, Kenya, 2016; ISBN 978-92-1-132708-3.
3. Salbitano, F.; Borelli, S.; Conigliaro, M.; Chen, Y. *Guidelines on Urban and Peri-Urban Forestry*; FAO Forest; Food and Agriculture Organization of the United Nations: Rome, Italy, 2016; ISBN 9789251094426.
4. Grote, R.; Samson, R.; Alonso, R.; Amorim, J.H.; Cariñanos, P.; Churkina, G.; Fares, S.; Thiec, D.L.; Niinemets, Ü.; Mikkelsen, T.N.; et al. Functional traits of urban trees: Air pollution mitigation potential. *Front. Ecol. Environ.* **2016**, *14*, 543–550. [CrossRef]
5. Churkina, G.; Grote, R.; Butler, T.M.; Lawrence, M. Natural selection? Picking the right trees for urban greening. *Environ. Sci. Policy* **2015**, *47*, 12–17. [CrossRef]
6. Maes, J.; Egoh, B.; Willemen, L.; Liquete, C.; Vihervaara, P.; Schägner, J.P.; Grizzetti, B.; Drakou, E.G.; Notte, A.L.; Zulian, G.; et al. Mapping ecosystem services for policy support and decision making in the European Union. *Ecosyst. Serv.* **2012**, *1*, 31–39. [CrossRef]
7. Endreny, T.; Santagata, R.; Perna, A.; De Stefano, C.; Rallo, R.F.; Ulgiati, S. Implementing and managing urban forests: A much needed conservation strategy to increase ecosystem services and urban wellbeing. *Ecol. Modell.* **2017**, *360*, 328–335. [CrossRef]
8. Nowak, D.; Crane, D. The urban forest effects (UFORE) model: Quantifying urban forest structure and function. In *Integrated Tools for Natural Resources Inventories in the 21st Century*; Hansen, M., Burk, T., Eds.; U.S. Department of Agriculture, Forest Service, North Central Forest Experiment Station: Saint Paul, MN, USA, 1998; pp. 714–720.

9. Hirabayashi, S.; Kroll, C.N.; Nowak, D.J. Development of a distributed air pollutant dry deposition modeling framework. *Environ. Pollut.* **2012**, *171*, 9–17. [CrossRef] [PubMed]
10. Nowak, D.J.; Hoehn, R.E.; Bodine, A.R.; Greenfield, E.J.; O'Neil-Dunne, J. Urban forest structure, ecosystem services and change in Syracuse, NY. *Urban Ecosyst.* **2013**, *19*, 1–23. [CrossRef]
11. Russo, A.; Escobedo, F.J.; Zerbe, S. Quantifying the local-scale ecosystem services provided by urban treed streetscapes in Bolzano, Italy. *AIMS Environ. Sci.* **2016**, *3*, 58–76. [CrossRef]
12. Selmi, W.; Weber, C.; Rivière, E.; Blond, N.; Mehdi, L.; Nowak, D. Air pollution removal by trees in public green spaces in Strasbourg city, France. *Urban For. Urban Green.* **2016**, *17*, 192–201. [CrossRef]
13. Westfall, J. Spatial-scale considerations for a large-area forest inventory regression model. *Forestry* **2015**, *88*, 267–274. [CrossRef]
14. Nowak, D.J.; Walton, J.; Stevens, J.C.; Crane, D.E.; Hoehn, R.E. Effect of Plot and Sample Size on Timing and Precision of Urban Forest Assessments METHODS Effect of Plot Size on Data Collection Time and Total Population Estimate Precision. *Arboric. Urban For.* **2008**, *34*, 386–390.
15. Bottalico, F.; Chirici, G.; Giannetti, F.; De Marco, A.; Nocentini, S.; Paoletti, E.; Salbitano, F.; Sanesi, G.; Serenelli, C.; Travaglini, D. Air Pollution Removal by Green Infrastructures and Urban Forests in the City of Florence. *Agric. Agric. Sci. Procedia* **2016**, *8*, 243–251. [CrossRef]
16. Manes, F.; Marando, F.; Capotorti, G.; Blasi, C.; Salvatori, E.; Fusaro, L.; Ciancarella, L.; Mircea, M.; Marchetti, M.; Chirici, G.; et al. Regulating Ecosystem Services of forests in ten Italian metropolitan Cities: Air quality improvement by PM_{10} and O_3 removal. *Ecol. Indic.* **2016**, *67*, 425–440. [CrossRef]
17. Marando, F.; Salvatori, E.; Fusaro, L.; Manes, F. Removal of PM_{10} by forests as a nature-based solution for air quality improvement in the Metropolitan city of Rome. *Forests* **2016**, *7*, 150. [CrossRef]
18. Fusaro, L.; Marando, F.; Sebastiani, A.; Capotorti, G.; Blasi, C.; Copiz, R.; Congedo, L.; Munafò, M.; Ciancarella, L.; Manes, F. Mapping and Assessment of PM_{10} and O_3 Removal by Woody Vegetation at Urban and Regional Level. *Remote Sens.* **2017**, *9*, 791. [CrossRef]
19. Bechtold, W.A. Crown position and light exposure classification-an alternative to field-assigned crown class. *North. J. Appl. For.* **2003**, *20*, 154–160.
20. Alonzo, M.; Bookhagen, B.; Roberts, D.A. Urban tree species mapping using hyperspectral and LiDAR data fusion. *Remote Sens. Environ.* **2014**, *148*, 70–83. [CrossRef]
21. Alonzo, M.; McFadden, J.P.; Nowak, D.J.; Roberts, D.A. Mapping urban forest structure and function using hyperspectral imagery and LiDAR data. *Urban For. Urban Green.* **2016**, *17*, 135–147. [CrossRef]
22. Parmehr, E.G.; Amati, M.; Taylor, E.J.; Livesley, S.J. Estimation of urban tree canopy cover using random point sampling and remote sensing methods. *Urban For. Urban Green.* **2016**, *20*, 160–171. [CrossRef]
23. Shojanoori, R.; Shafri, H.Z.M. Review on the Use of Remote Sensing for Urban Forest Monitoring. *Arboric. Urban For.* **2016**, *42*, 400–417.
24. Yang, J.; Chang, Y.M.; Yan, P.B. Ranking the suitability of common urban tree species for controlling $PM_{2.5}$ pollution. *Atmos. Pollut. Res.* **2015**, *6*, 267–277. [CrossRef]
25. Fassnacht, F.E.; Latifi, H.; Stereńczak, K.; Modzelewska, A.; Lefsky, M.; Waser, L.T.; Straub, C.; Ghosh, A. Review of studies on tree species classification from remotely sensed data. *Remote Sens. Environ.* **2016**, *186*, 64–87. [CrossRef]
26. Berland, A.; Lange, D.A. Google Street View shows promise for virtual street tree surveys. *Urban For. Urban Green.* **2017**, *21*, 11–15. [CrossRef]
27. Tanhuanpää, T.; Vastaranta, M.; Kankare, V.; Holopainen, M.; Hyyppä, J.; Hyyppä, H.; Alho, P.; Raisio, J. Mapping of urban roadside trees—A case study in the tree register update process in Helsinki City. *Urban For. Urban Green.* **2014**, *13*, 562–570. [CrossRef]
28. Bella, I.E. A new competition model for individual trees. *For. Sci.* **1971**, *17*, 364–372.
29. Korol, R.L.; Running, S.W.; Milner, K.S. Incorporating intertree competition into an ecosystem model. *Can. J. For. Res.* **1995**, *25*, 413–424. [CrossRef]
30. Fox, J.C.; Bi, H.; Ades, P.K. Spatial dependence and individual-tree growth models. II. Modelling spatial dependence. *For. Ecol. Manag.* **2007**, *245*, 20–30. [CrossRef]
31. Pretzsch, H.; Biber, P.; Dursky, J. The single tree-based stand simulator SILVA: Construction, application and evaluation. *For. Ecol. Manag.* **2002**, *162*, 3–21. [CrossRef]
32. Nowak, D.J.; Crane, D.E.; Stevens, J.C.; Hoehn, R.E.; Walton, J.T.; Bond, J. A Ground-Based Method of Assessing Urban Forest Structure and Ecosystem Services. *Arboric. Urban For.* **2008**, *34*, 347–358. [CrossRef]

33. USDA Forest Service. *i-Tree Eco User's Manual v 6.0*; U.S. Forest Service Northern Research Station (NRS): Washington, DC, USA, 2016.
34. Nowak, D.I. Estimating Leaf Area and Leaf Biomass of Open-Grown Deciduous Urban Trees. *For. Sci.* **1996**, *42*, 504–507.
35. Nowak, D.J.; Crane, D.E. Carbon storage and sequestration by urban trees in the USA. *Environ. Pollut.* **2002**, *116*, 381–389. [CrossRef]
36. BMEL Federal Ministry of Food and Agriculture. *The Forests in Germany: Selected Results of the Third National Forest Inventory*; Federal Ministry of Food and Agriculture: Berlin, Germany, 2015.
37. Hirabayashi, S. *Air Pollutant Removals, Biogenic Emissions and Hydrologic Estimates for i-Tree Applications*; United States Forest Service: Syracuse, NY, USA, 2016.
38. Nowak, D.J. Atmospheric Carbon Dioxide Reduction by Chicago's urban forest. In *Chicago's Urban Forest Ecosystem: Results of the Chicago Urban Forest Climate Project*; McPherson, E.G., Nowak, D.J., Eds.; US Department of Agriculture, Forest Service, Northeastern Forest Experiment Station: Radnor, PA, USA, 1994; pp. 83–94.
39. Cairns, M.A.; Brown, S.; Helmer, E.H.; Baumgardner, G.A. Root biomass allocation in the world's upland forests. *Oecologia* **1997**, *111*, 1–11. [CrossRef] [PubMed]
40. Hirabayashi, S.; Kroll, C.N.; Nowak, D.J. Component-based development and sensitivity analyses of an air pollutant dry deposition model. *Environ. Model. Softw.* **2011**, *26*, 804–816. [CrossRef]
41. Nowak, D.J.; Crane, D.E.; Stevens, J.C. Air pollution removal by urban trees and shrubs in the United States. *Urban For. Urban Green.* **2006**, *4*, 115–123. [CrossRef]
42. Nowak, D.J.; Hirabayashi, S.; Bodine, A.; Hoehn, R. Modeled $PM_{2.5}$ removal by trees in ten US cities and associated health effects. *Environ. Pollut.* **2013**, *178*, 395–402. [CrossRef] [PubMed]
43. Nowak, D.J.; Hirabayashi, S.; Bodine, A.; Greenfield, E. Tree and forest effects on air quality and human health in the United States. *Environ. Pollut.* **2014**, *193*, 119–129. [CrossRef] [PubMed]
44. Baldocchi, D.D.; Hicks, B.B.; Camara, P. A canopy stomatal resistance model for gaseous deposition to vegetated surfaces. *Atmos. Environ.* **1987**, *21*, 91–101. [CrossRef]
45. Pederson, J.R.; Massman, W.J.; Mahrt, L.; Delany, A.; Oncley, S.; Hartog, G.D.; Neumann, H.H.; Mickle, R.E.; Shaw, R.H.; Paw U, K.T.; et al. California ozone deposition experiment: Methods, results, and opportunities. *Atmos. Environ.* **1995**, *29*, 3115–3132. [CrossRef]
46. Hirabayashi, S.; Kroll, C.N.; Nowak, D.J. *i-Tree Eco Dry Deposition Model Descriptions*; United States Forest Service: Syracuse, NY, USA, 2015.
47. Baldocchi, D. A Multi-layer model for estimating sulfur dioxide deposition to a deciduous oak forest canopy. *Atmos. Environ.* **1988**, *22*, 869–884. [CrossRef]
48. Hosker, R.P.; Lindberg, S.E. Review: Atmospheric deposition and plant assimilation of gases and particles. *Atmos. Environ.* **1982**, *16*, 889–910. [CrossRef]
49. Wesley, M.L. Parametrization of surface resistance to gaseous dry deposition in regional-scale numerical model. *Atmos. Environ.* **1989**, *23*, 1293–1304. [CrossRef]
50. Taylor, G.E.; Hanson, P.J.; Baldocchi, D.D. Pollutant deposition to individual leaves and plant canopies: Sites of regulation and relationship to injury. In *Assessment of Crop Loss from Air Pollution*; Heck, W.W., Taylor, O.C., Tingey, D.T., Eds.; Springer: Dordrecht, The Netherlands, 1988; pp. 227–257.
51. Lovett, G.M. Atmospheric deposition of nutrients and pollutants in North America: An ecological perspective. *Ecol. Appl.* **1994**, *4*, 629–650. [CrossRef]
52. Bidwell, R.G.S.; Fraser, D.E. Carbon monoxide uptake and metabolism by leaves. *Can. J. Bot.* **1972**, *50*, 1435–1439. [CrossRef]
53. Guenther, A.B.; Zimmerman, P.R.; Harley, P.C.; Monson, R.K. Isoprene and Monoterpene Emission Rate Variability' Model Evaluations and Sensitivity Analyses. *J. Geophys. Res.* **1993**, *98617*, 609–612. [CrossRef]
54. Geron, C.D.; Guenther, A.B.; Pierce, T.E. An improved model for estimating emissions of volatile organic compounds from forests in the eastern United States. *J. Geophys. Res.* **1994**, *99*, 12773. [CrossRef]
55. Hirabayashi, S. *i-Tree Eco Biogenic Emissions Model Descriptions*; United States Forest Service: Syracuse, NY, USA, 2012.
56. Nowak, D.J.; Crane, D.E.; Stevens, J.C.; Ibarra, M. *Brooklyn's Urban Forest*; U.S. Department of Agriculture, Forest Service, Northeastern Research Station: Newtown Square, PA, USA, 2002.
57. Monteith, J.L.; Unsworth, M.H. *Principles of Environmental Physics*, 2nd ed.; Edward Arnold: London, UK, 1990.

58. Rogers, K.; Sacre, K.; Goodenough, J.; Doick, K. *Valuing London's Urban Forest*; Treeconomics: London, UK, 2015; ISBN 9780957137110.
59. Grantz, D.A.; Gunn, S.; Vu, H.B. O_3 impacts on plant development: A meta-analysis of root/shoot allocation and growth. *Plant Cell Environ.* **2006**, *29*, 1193–1209. [CrossRef] [PubMed]
60. Landolt, W.; Bühlmann, U.; Bleuler, P.; Bucher, J.B. Ozone exposure–response relationships for biomass and root/shoot ratio of beech (*Fagus sylvatica*), ash (*Fraxinus excelsior*), Norway spruce (*Picea abies*) and Scots pine (Pinus sylvestris). *Environ. Pollut.* **2000**, *109*, 473–478. [CrossRef]
61. Moser, A.; Rötzer, T.; Pauleit, S.; Pretzsch, H. The Urban Environment Can Modify Drought Stress of Small-Leaved Lime (*Tilia cordata* Mill.) and Black Locust (*Robinia pseudoacacia* L.). *Forests* **2016**, *7*, 71. [CrossRef]
62. Pretzsch, H.; Biber, P.; Uhl, E.; Dahlhausen, J.; Schütze, G.; Perkins, D.; Rötzer, T.; Caldentey, J.; Koike, T.; van Con, T.; et al. Climate change accelerates growth of urban trees in metropolises worldwide. *Sci. Rep.* **2017**, *7*, 15403. [CrossRef] [PubMed]
63. Dahlhausen, J.; Rötzer, T.; Biber, P.; Uhl, E.; Pretzsch, H. Urban climate modifies tree growth in Berlin. *Int. J. Biometeorol.* **2017**, 1–14. [CrossRef] [PubMed]
64. McHale, M.R.; Burke, I.C.; Lefsky, M.A.; Peper, P.J.; McPherson, E.G. Urban forest biomass estimates: Is it important to use allometric relationships developed specifically for urban trees? *Urban Ecosyst.* **2009**, *12*, 95–113. [CrossRef]
65. Russo, A.; Escobedo, F.J.; Timilsina, N.; Schmitt, A.O.; Varela, S.; Zerbe, S. Assessing urban tree carbon storage and sequestration in Bolzano, Italy. *Int. J. Biodivers. Sci. Ecosyst. Serv. Manag.* **2014**, *10*, 54–70. [CrossRef]
66. König, G.; Brunda, M.; Puxbaum, H.; Hewitt, C.N.; Duckham, S.C.; Rudolph, J. Relative contribution of oxygenated hydrocarbons to the total biogenic VOC emissions of selected mid-European agricultural and natural plant species. *Atmos. Environ.* **1995**, *29*, 861–874. [CrossRef]
67. Moukhtar, S.; Bessagnet, B.; Rouil, L.; Simon, V. Monoterpene emissions from Beech (*Fagus sylvatica*) in a French forest and impact on secondary pollutants formation at regional scale. *Atmos. Environ.* **2005**, *39*, 3535–3547. [CrossRef]
68. Aydin, Y.M.; Yaman, B.; Koca, H.; Dasdemir, O.; Kara, M.; Altiok, H.; Dumanoglu, Y.; Bayram, A.; Tolunay, D.; Odabasi, M.; et al. Biogenic volatile organic compound (BVOC) emissions from forested areas in Turkey: Determination of specific emission rates for thirty-one tree species. *Sci. Total Environ.* **2014**, *490*, 239–253. [CrossRef] [PubMed]
69. Papiez, M.R.; Potosnak, M.J.; Goliff, W.S.; Guenther, A.B. The impacts of reactive terpene emissions from plants on air quality in Las Vegas, Nevada. *Atmos. Environ.* **2009**, *43*, 4109–4123. [CrossRef]
70. Tiwary, A.; Namdeo, A.; Fuentes, J.; Dore, A.; Hu, X.; Bell, M. Systems scale assessment of the sustainability implications of emerging green initiatives. *Environ. Pollut.* **2013**, *183*, 213–223. [CrossRef] [PubMed]
71. Curtis, A.J.; Helmig, D.; Baroch, C.; Daly, R.; Davis, S. Biogenic volatile organic compound emissions from nine tree species used in an urban tree-planting program. *Atmos. Environ.* **2014**, *95*, 634–643. [CrossRef]
72. Dunn-Johnston, K.A.; Kreuzwieser, J.; Hirabayashi, S.; Plant, L.; Rennenberg, H.; Schmidt, S. Isoprene Emission Factors for Subtropical Street Trees for Regional Air Quality Modeling. *J. Environ. Qual.* **2016**, *45*, 234–243. [CrossRef] [PubMed]
73. Monson, R.K.; Grote, R.; Niinemets, Ü.; Schnitzler, J.P. Modeling the isoprene emission rate from leaves. *New Phytol.* **2012**, *195*, 541–559. [CrossRef] [PubMed]
74. Ghirardo, A.; Xie, J.; Zheng, X.; Wang, Y.; Grote, R.; Block, K.; Wildt, J.; Mentel, T.; Kiendler-Scharr, A.; Hallquist, M.; et al. Urban stress-induced biogenic VOC emissions and SOA-forming potentials in Beijing. *Atmos. Chem. Phys.* **2016**, *16*, 2901–2920. [CrossRef]
75. Grote, R.; Lavoir, A.V.; Rambal, S.; Staudt, M.; Zimmer, I.; Schnitzler, J.P. Modelling the drought impact on monoterpene fluxes from an evergreen Mediterranean forest canopy. *Oecologia* **2009**, *160*, 213–223. [CrossRef] [PubMed]
76. Bourtsoukidis, E.; Kawaletz, H.; Radacki, D.; Schütz, S.; Hakola, H.; Hellén, H.; Noe, S.; Mölder, I.; Ammer, C.; Bonn, B. Impact of flooding and drought conditions on the emission of volatile organic compounds of *Quercus robur* and *Prunus serotina*. *Trees* **2014**, *28*, 193–204. [CrossRef]
77. Derwent, R.G.; Jenkin, M.E.; Saunders, S.M. Photochemical ozone creation potentials for a large number of reactive hydrocarbons under European conditions. *Atmos. Environ.* **1996**, *30*, 181–199. [CrossRef]

78. Cabaraban, M.T.I.; Kroll, C.N.; Hirabayashi, S.; Nowak, D.J. Modeling of air pollutant removal by dry deposition to urban trees using a WRF/CMAQ/i-Tree Eco coupled system. *Environ. Pollut.* **2013**, *176*, 123–133. [CrossRef] [PubMed]
79. McPherson, E.; Peper, P. Urban tree growth modeling. *Arboric. Urban For.* **2012**, *38*, 172–180.
80. Fares, S.; Savi, F.; Muller, J.; Matteucci, G.; Paoletti, E. Simultaneous measurements of above and below canopy ozone fluxes help partitioning ozone deposition between its various sinks in a Mediterranean Oak Forest. *Agric. For. Meteorol.* **2014**, *198*, 181–191. [CrossRef]
81. Morani, A.; Nowak, D.; Hirabayashi, S.; Guidolotti, G.; Medori, M.; Muzzini, V.; Fares, S.; Mugnozza, G.S.; Calfapietra, C. Comparing i-Tree modeled ozone deposition with field measurements in a periurban Mediterranean forest. *Environ. Pollut.* **2014**, *195*, 202–209. [CrossRef] [PubMed]
82. Beckett, K.P.; Freer-Smith, P.H.; Taylor, G. Particulate pollution capture by urban trees: Effect of species and windspeed. *Glob. Chang. Biol.* **2000**, *6*, 995–1003. [CrossRef]
83. Kardel, F.; Wuyts, K.; Babanezhad, M.; Wuytack, T.; Adriaenssens, S.; Samson, R. Tree leaf wettability as passive bio-indicator of urban habitat quality. *Environ. Exp. Bot.* **2012**, *75*, 277–285. [CrossRef]
84. Hofman, J.; Wuyts, K.; Van Wittenberghe, S.; Samson, R. On the temporal variation of leaf magnetic parameters: Seasonal accumulation of leaf-deposited and leaf-encapsulated particles of a roadside tree crown. *Sci. Total Environ.* **2014**, *493*, 766–772. [CrossRef] [PubMed]
85. Amorim, J.H.; Rodrigues, V.; Tavares, R.; Valente, J.; Borrego, C. CFD modelling of the aerodynamic effect of trees on urban air pollution dispersion. *Sci. Total Environ.* **2013**, *461–462*, 541–551. [CrossRef] [PubMed]
86. Harris, T.B.; Manning, W.J. Nitrogen dioxide and ozone levels in urban tree canopies. *Environ. Pollut.* **2010**, *158*, 2384–2386. [CrossRef] [PubMed]
87. Pihlatie, M.; Rannik, Ü.; Haapanala, S.; Peltola, O.; Shurpali, N.; Martikainen, P.J.; Lind, S.; Hyvönen, N.; Virkajärvi, P.; Zahniser, M.; et al. Seasonal and diurnal variation in CO fluxes from an agricultural bioenergy crop. *Biogeosciences* **2016**, *13*, 5471–5485. [CrossRef]
88. Sanhueza, E.; Dong, Y.; Scharffe, D.; Lobert, J.M.; Crutzen, P.J.; Dong, Y.; Scharffe, D.; Lobert, J.M.; Carbon, P.J.C. Carbon monoxide uptake by temperate forest soils: The effects of leaves and humus layers. *Tellus B Chem. Phys. Meteorol.* **1998**, *50*, 51–58. [CrossRef]

© 2018 by the authors. Licensee MDPI, Basel, Switzerland. This article is an open access article distributed under the terms and conditions of the Creative Commons Attribution (CC BY) license (http://creativecommons.org/licenses/by/4.0/).

Article

How Do *Tilia cordata* Greenspire Trees Cope with Drought Stress Regarding Their Biomass Allocation and Ecosystem Services?

Chi Zhang [1],*, Laura Myrtiá Faní Stratopoulos [2], Hans Pretzsch [1] and Thomas Rötzer [1]

[1] Chair for Forest Growth and Yield Science, School of Life Sciences, Technische Universität München, Hans-Carl-von-Carlowitz-Platz 2, 85354 Freising, Germany
[2] Department of Landscape Architecture, Weihenstephan-Triesdorf University of Applied Sciences, Weihenstephaner Berg 17, 85354 Freising, Germany
* Correspondence: forestrychi.zhang@tum.de; Tel.: +49(0)8161-71-4719

Received: 29 June 2019; Accepted: 8 August 2019; Published: 9 August 2019

Abstract: In the context of climate change, drought is likely to become more frequent and more severe in urban areas. Urban trees are considered to play an important role in fixing carbon, improving air quality, reducing noise and providing other ecosystem services. However, data on the response of urban trees to climate change, particularly to drought, as well as the relationship between their below- and above-ground processes in this context, are still limited, which prevents a comprehensive understanding of the role of urban trees in ameliorating some of the adverse effects of climate change and their ability to cope with it. To investigate whole-plant responses to water shortages, we studied the growth of *Tilia cordata* Greenspire, a commonly planted urban tree, including development of its roots and stem diameter, leaf parameters and the harvested biomass. Our results showed that this cultivar was susceptible to drought and had reduced biomass in all three compartments: branch (30.7%), stem (16.7%) and coarse roots (45.2%). The decrease in the root:shoot ratio under drought suggested that more carbon was invested in the above-ground biomass. The development of fine roots and the loss of coarse root biomass showed that *T. cordata* Greenspire prioritised the growth of fine roots within the root system. The *CityTree* model's simulation showed that the ability of this cultivar to provide ecosystem services, including cooling and CO_2 fixation, was severely reduced. For use in harsh and dry urban environments, we recommend that urban managers take into account the capacity of trees to adapt to drought stress and provide sufficient rooting space, especially vertically, to help trees cope with drought.

Keywords: biomass allocation; drought; ecosystem services; root:shoot ratio; urban trees

1. Introduction

In the future, drought is projected to occur more frequently under the warmer conditions associated with the progression of climate change [1]. Additionally, in association with increasing urbanisation, extreme heat events have become more prevalent than in previous decades, which reduce environmental quality [2–4]. Urban trees can mitigate environmental degradation by storing carbon, purifying the air, reducing storm water and providing other ecosystem services [5–7]. Urban trees can also be a key component in the adaptation of cities to climate change [8]. Hence, in recent decades, the use of urban trees for various purposes has drawn increasing interest from researchers [9]. For example, Nowak et al. [10] studied how urban trees can filter pollutants that have adverse effects on human health. Konarska et al. [11] quantified the magnitude of daytime and night-time transpiration of common urban tree species. In addition, Velasco et al. [12] reported that carbon sequestration depends on the characteristics of urban trees, and Pretzsch et al. [13] proposed that urban climates can either accelerate or decelerate tree growth.

Water availability is considered to be the most important determinant of tree growth [14,15], and a higher frequency of drought events will expose urban trees to more restrictive growth conditions. To improve the quality of ecosystem services, how different urban tree species cope with drought stress has been commonly discussed [16]. Rötzer et al. [17] pointed out that stress caused by a water deficit could reduce photosynthetic productivity and tree growth. Moser et al. [18] found marked growth reductions during drought periods and subsequent rapid recovery in *Robinia pseudoacacia* L. Furthermore, Stratópoulos et al. [19] proposed that trees from dry regions, such as *Ostrya carpinifolia* Scop. and *Tilia tomentosa* Moench 'Brabant', show a high tolerance to drought stress; however, this tolerance can come at the expense of above-ground biomass production.

The adaptation of urban trees to drought has been a recent topic of research, but there is still a knowledge gap regarding carbon allocation as well as ecosystem services [20]. Carbon allocation between below- and above-ground biomasses is one of the key parameters towards understanding tree survival, especially under the global change accompanied by urbanisation [21]. Data on the relationship between below- and above-ground processes is limited, which may result in scarce information of the ecophysiology and hydric behaviour of urban trees under climate change [22–24]. The development of below-ground biomass has also been rarely studied because of multiple factors such as soil compaction, limited root volume and harsh paved environments [25–27]. Furthermore, urban trees can make an enormous contribution to mitigating the urban heat islands by providing a cooling effect [28,29], which is considered a feasible option for adapting to climate change [30]. Hence, information regarding how urban trees allocate their biomass and provide ecosystem services such as cooling and carbon fixation is crucial for city planners to implement appropriate management practices [31].

In some previous studies, only allometric equations were used to estimate such allocation because of the expensive and time-consuming harvesting process [32]. This has led to a lack of precise information on carbon allocation, and hence researchers have been prevented from establishing appropriate strategies for managing urban trees. Against this background, in this study, we selected *Tilia cordata* Mill. 'Greenspire', a cultivar widely planted in Central European cities, to analyse tree growth and carbon allocation under undisturbed growing conditions (control) and under extreme drought. Combined with a harvesting campaign, we applied the urban tree model *CityTree* to simulate biomass development as well as ecosystem services. The following research questions are addressed in this work: (1) How does *T. cordata* Greenspire respond to extreme drought in terms of growth and carbon allocation? (2) What is the cultivar's strategy within the root system? (3) Are its ecosystem services severely affected under drought?

2. Materials and Methods

2.1. Study Site and Drought Experiment in 2017

This study was performed at the municipal nursery of Munich (48°08′05″ N, 11°28′47″ E, 534 m a.s.l.), the major city in the southeast of Germany under the marine west coast climate. With the predominant soil types ranging from moderate sandy loam to strong loamy sand, the 42 ha nursery shows very little variation in terms of microclimatic conditions. The measurements in this study were performed from April 12 to November 12, 2017. Eight young individuals of *T. cordata* Greenspire with diameters of 5–6 cm at a height of 1 m were selected, four of which were set as a 'control' group and the other four as a 'dry-treatment' group. Each tree was more than 2 m away from the others to avoid the mutual effect.

For the drought-stress experiment, we used experimental settings in which conditions of an enduring drought event were simulated to investigate the tree growth responses. The rooting spaces of the four trees in the dry-treatment group were covered by a rainfall exclusion roof (RER) from May to November in 2017, whereas the four trees in the control group were exposed to normal weather conditions. The RER was made up of several waterproof tents (2 × 3 m) combined with nylon ropes and installed at a height of 1 m, sloping to the ground in the north and south directions at 3 m from the trees (Figure 1).

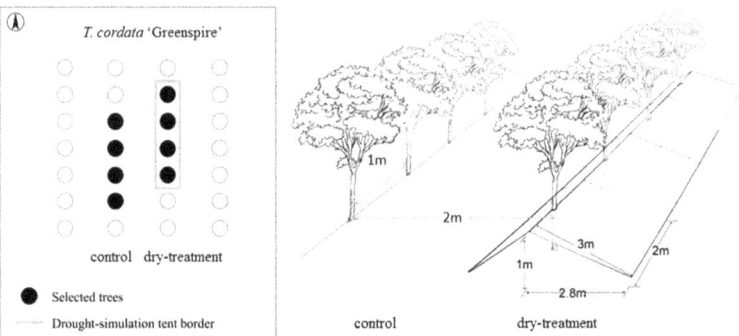

Figure 1. Schematic structure of the drought experiment. The rooting zones of trees on the right were covered by rainfall exclusion constructions ('dry-treatment'), whereas trees on the left were exposed to normal weather conditions ('control').

2.2. Measurement of Climate Variables and Soil Moisture

Climate variables, including temperature and precipitation, were sampled every 10 min with a weather station (Davis Vantage Pro2; Davis Instruments, Hayward, CA, USA) located at an unshaded site approximately 200 m away from the experimental plots. 2017 was a warm year at the site, with an mean temperature of 10.3 °C and a total precipitation of 887 mm. The spring of 2017 was warm, sunny and dry. After a strong return to cold weather in April, the temperature started to increase at the beginning of May and remained exceptionally high, in that the average temperature was 19.6 °C from July to August. In this period, monthly precipitation levels of 101 and 123 mm were similar to the long-term records (within 1981–2010) of 122 and 115 mm. The temperature showed a declining trend in early September, and night frost started in the middle of November (DWD, 2018) (see Figure 2).

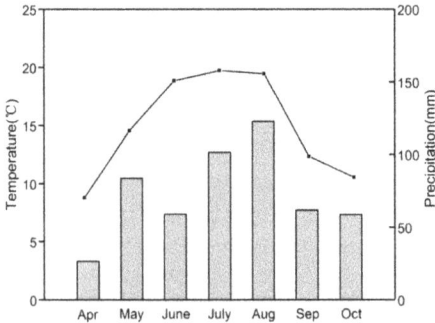

Figure 2. Monthly totals of precipitation (mm; bar graph) and monthly average temperature (°C; dotted line) for the study site from April to October 2017.

We measured the volumetric soil water content (VWC; Vol.%) with a portable tool (UMP-1; UGT, Müncheberg, Germany) at the rooting zones of all the trees four times in the summer of 2017, with the goal of comparing the moisture differences between the control and the dry-treatment groups (see Table 1). The values of VWC between the control and the dry-treatment groups were significantly different for all the four times ($p < 0.01$), which proved the success of the drought simulation.

Table 1. Volumetric soil water content (Vol.%; VWC) at 10 cm soil depth for the control and dry treatment groups of *T. cordata* Greenspire four times in 2017.

Group	n	Date			
		July 12	July 31	Aug 16	Sep 7
control	4	16.5 ± 1.3	19.6 ± 0.7	25.3 ± 4.6	21.3 ± 1.7
dry-treatment	4	12.0 ± 1.2	12.1 ± 2.0	12.8 ± 1.0	13.4 ± 1.6

2.3. Measurement of Above- and Below-Ground Biomasses

The stem diameter at a height of 1 m was measured using a digital caliper at the beginning of each month from April to November 2017. Measurements in two perpendicular directions (N–S and E–W) were performed and averaged.

Fine root coring campaigns were launched for all trees in May, September and November: that is, at the beginning (pre-drought), in the middle and at the end of the growing season. A pre-test coring campaign showed that the range of the root system was similar to a cylinder, with a diameter of 70 cm and a height of 35 cm. Therefore, during every coring campaign, four soil cores were collected for every individual tree: two at a distance of 15 cm from the trunk and two at a distance of 30 cm. The soil was sampled down to a depth of 30 cm using a soil auger with a length of 30 cm and a radius of 3 cm. Each sample was divided into three horizons: soil depths of 0–10 cm (upper layer), 10–20 cm (middle layer) and 20–30 cm (deep layer). Fine roots (< 2 cm) were filtered using sieves (2-mm mesh size) and separated by forceps in the laboratory. Then, the samples were washed and dried in an oven at 65°C for 72 h. Finally, all the samples were weighed using a balance with an accuracy of four decimal places to obtain the dry weight. The fine root biomass at different depths was calculated using the dry weight divided by the cross-sectional area of the auger.

In November, a harvest campaign was launched in which all the trees were excavated with a tree digger and divided into three parts: branch, stem and coarse roots. All these compartments were dried at 65 °C for 72 h and weighed using a balance with an accuracy of up to four decimal places to obtain the dry weight. Before drying, the root systems were washed to remove attached soil and stones. An image analysis process including taking high-resolution photos from five views (N, S, W, E and top) was applied to measure the root architecture with the help of the software 'Root System Analyzer' (RSA, University of Vienna, Austria). On the basis of a graphical representation of the skeletonised image of the root system as well as segmentation algorithms, RSA was used to describe some of the root traits, including root nodes, width and depth.

The leaf area index ($m^2\ m^{-2}$; *LAI*) was determined using hemispherical photographs (Nikon Coolpix P5100 camera with a fisheye lens and Mid-OMount) and analysed with the program WinSCANOPY (Régent Instruments Inc., Quebec, Canada). Data acquisition was performed under conditions of a uniformly overcast sky in the middle of June, shortly after implementation of the drought experiment. Sufficient numbers of points (7–11 x, z pairs) were measured and input into the software FV2200 (LICOR Biosciences, Lincoln, NE) to compute the projected crown area (*PCA*) of each tree. Combined with the specific leaf area (*SLA*), the leaf biomass was calculated as follows:

$$Biomass_{leaf} = \frac{LAI \times PCA}{SLA} \quad (1)$$

Therefore, the below-ground biomass was the sum of the fine and coarse root biomasses, and the above-ground biomass was the sum of the branch, stem and leaf biomasses.

2.4. Simulation of Biomass and Ecosystem Services

A process-based model was used to simulate the wood biomass of *T. cordata* Greenspire and calculate its ecosystem services [33,34]. On the basis of the basic measurements of trees (e.g., DBH (diameter at breast height) and tree height), climate and soil data, this model consists of seven modules to calculate tree growth and ecosystem services such as CO_2 fixation, evapotranspirational cooling and shading. The core function of the simulation of a tree's net assimilation is as follows:

$$A = d \times \left[(J_p + J_r) - \sqrt{(J_p + J_r)^2) - 4 \times \theta \times J_p \times J_r} \right] / 2 \times \theta \qquad (2)$$

where A is gross assimilation (g C m^{-2} d^{-1}), d is mean day length of the month, J_p is reaction of photosynthesis on absorbed photosynthetic radiation (g C m^{-2} h^{-1}), J_r is the Rubisco-limited rate of photosynthesis (g C m^{-2} h^{-1}), and θ is the form factor (= 0.7).

J_p is a function of the photosynthetic active radiation (PAR) and the efficiency of carbon fixation per absorbed PAR, which can be calculated on the basis of the intrinsic quantum efficiency for CO_2 uptake, the partial pressure of internal CO_2, the CO_2 compensation point, the influence of temperature on the efficiency and a species-dependent adjustment function for tree age. The Rubisco-limited rate of photosynthesis J_r can be estimated by the maximum catalytic Rubisco capacity, the maximum day length, the Michaelis–Menten constant of CO_2 depending on temperature, the inhibition constant of O_2 against CO_2 (temperature-dependent) and the O_2 concentration.

Net assimilation A_N is calculated as follows:

$$A_N = A - R_d \qquad (3)$$

where R_d is the product of the maximum catalytic Rubisco capacity and the ratio of the maximum catalytic Rubisco capacity and the respiration cost. A fixed share of 50% of the net assimilation is assumed for growth and maintenance respiration [35]. The tree growth represented the fixation of carbon, and the fixation of CO_2 was calculated based on fixation of carbon and the relative molecular mass.

For the ecosystem service of evapotranspiration, the central water balance equation from the water balance module was as:

$$prec - int - et_a - ro - \Delta\varphi = 0 \qquad (4)$$

With $prec$ = precipitation (mm), int = interception (mm), et_a = actual evapotranspiration (mm), ro = runoff (mm), $\Delta\varphi$ = change of the soil content (mm).

Within the module cooling the energy needed for the transition of water from liquid to gaseous phase was calculated based on the CPA (crown projected area) and the transpiration et_a sum:

$$E_A = et_a \times CPA - (L_O \times -0.00242 \times temp) / f_{con} \qquad (5)$$

With E_A: energy released by a tree through transpiration (kWh tree^{-1}), L_O: energy needed for the transition of the 1 kg of water from the liquid to gaseous phase = 2.498 MJ (kgH$_2$O)$^{-1}$ and $temp$ = temperature in °C, f_{con}: conversion factor.

2.5. Statistical Analysis

The software package R [36] was used for statistical analysis. To investigate the difference between means, two-sampled t-test and analysis of variance (ANOVA) with Tukey's HSD (honestly significant difference) test were used. In all the cases the means were reported as significant when $p < 0.05$. Where necessary, data were log or power transformed in order to correct for data displaying heteroscedasticity.

3. Result

3.1. Stem Growth Under Drought

At the beginning of the growing season, the two groups had similar and slight decreases in diameter. From May to August, distinctly different increases were observed, with the control group exhibiting more rapid growth than the dry-treatment group ($p < 0.05$). Despite the similar patterns, in autumn, the dry-treatment group had a more obvious and severe decrease in the diameter increment (see Figure 3).

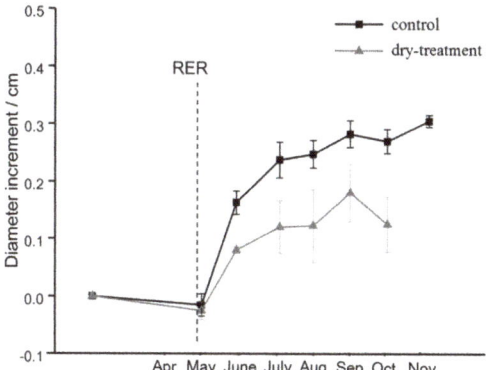

Figure 3. Stem diameter growth of the control and dry-treatment groups of *T. cordata* Greenspire measured at a height of 1 m from April to November 2017. The panel represents the mean value, and error bars indicate standard deviation. Due to an unexpected failure of the digital caliper, the data for the treatment group in November was missing.

3.2. Fine Root Development and Root Image Analysis

The total fine root biomasses of the control and dry-treatment groups were not significantly different in May, September and November ($p > 0.05$). However, we found different patterns from the control and treatment group. From May to September, the total fine root biomass in the control group had an obvious increase from 118.8 to 160.4 g m^{-2} while that decreased from 177.9 to 145.7 g m^{-2} in the treatment group ($p < 0.05$). From September to November, both the control and treatment groups had slight growths from 160.4 to 162.2 g m^{-2} and from 145.7 to 155.8 g m^{-2} ($p < 0.05$).

For more detailed information of the dynamics of the fine root biomass, we divided the fine roots into three layers. For the control group, except for a distinct growth for the deep layer (from 43.5 to 84.9 g m^{-2}), slight development was found for the other circumstances (see Figure 4). From May to September, however, the biomass in the upper and middle soil layers (0–10 and 10–20 cm) in the dry-treatment group decreased from 43.4 and 70.1 g m^{-2} to 10.0 and 29.9 g m^{-2}, whereas the fine roots in the deep layer still maintained distinct development from 64.4 to 105.8 g m^{-2}. Additionally in autumn, an adverse growth pattern was observed: only the deep fine root biomass decreased, whereas the upper and middle fine roots increased from 10.0 to 30.1 g m^{-2} and from 29.9 to 48.9 g m^{-2}, respectively.

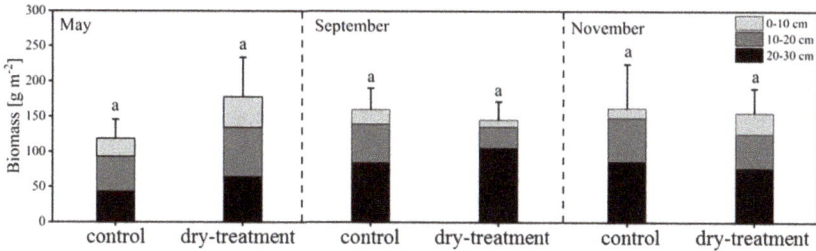

Figure 4. Development of fine root biomass from both the control and dry-treatment groups in May, September and November. Data are means of four soil cores for each individual tree for three vertical layers: 0–10, 10–20 and 20–30 cm. The letters indicate no significant differences ($p > 0.05$) between the control and dry-treatment groups for the total fine root biomass.

Although no significant difference was found between the control and dry-treatment groups ($p > 0.05$), the RSA software showed that the root system in the dry-treatment group exhibited decreases compared with the control group in terms of width, root number and node number. The numbers of root and node in the dry-treatment group were 1015 ± 155 and 804 ± 116, which were less than the control ones with the number of 1087 ± 182 and 841 ± 130. Besides maintaining the rooting depth, the root system in the dry-treatment group showed a slight decrease from 0.69 ± 0.04 to 0.65 ± 0.07 m in rooting width (see Figure 5).

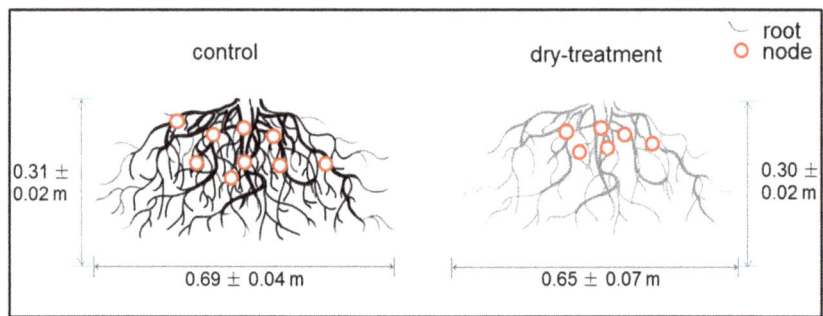

Figure 5. Two-dimensional illustration of the root systems in the control and dry-treatment groups. Red circles represent the rooting nodes.

3.3. Biomass Allocation and Root:Shoot Ratio

Generally, the stem biomass had the largest proportion of the tree, followed by the root system, and the branches had the lowest share (see Table 2). From the harvesting campaign, reductions of branch biomass (30.7%), stem biomass (16.7%) and coarse root biomass (45.2%) were observed. The biomasses of all three compartments in the dry-treatment group showed a distinct decrease in comparison to the control group, with the coarse roots in particular showing the largest difference ($P_{branch} < 0.01$, $P_{stem} < 0.05$ and $P_{coarse\ roots} < 0.001$).

Table 2. Biomasses of branch, stem and coarse roots of both the control and the dry-treatment groups from the harvesting campaign as well as leaf area index (LAI) and projected canopy area (PCA) measured in summer 2017.

Group	n	Wood Biomass			Leaf		
		Branch (g ± sd)	Stem (g ± sd)	Coarse Root (g ± sd)	LAI ($m^2\ m^{-2}$ ± sd)	PCA (m^2 ± sd)	SLA [1] ($m^2\ g^{-1}$)
control	4	445.8 ± 31.0	4089.3 ± 220.2	2440.3 ± 219.5	2.53 ± 0.25	0.58 ± 0.04	0.023
dry-treatment	4	308.5 ± 49.1	3407.0 ± 322.4	1338.7 ± 89.4	2.13 ± 0.24	0.58 ± 0.05	

In this study, we did not measure the SLA but used the same value of specific leaf area (SLA) from the literature for both the control and treatment group. [1] According to [34], the SLA of *T. cordata* in urban areas is 23.44 kg m^{-2}.

Three types of root:shoot ratio were calculated: (1) fine root:leaf biomass ratio, (2) coarse root:branch biomass ratio and (3) below-ground:above-ground biomass ratio. All the root:shoot ratios were significantly different ($p < 0.05$), with roots in the dry-treatment group constituting less of a proportion of the whole tree than in the control group (Figure 6). Among the ratios, the coarse root:branch ratio showed the greatest difference ($p < 0.01$).

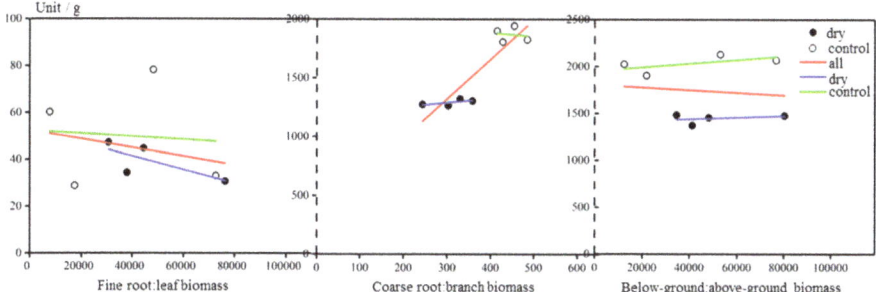

Figure 6. Three types of root:shoot ratio: fine root:leaf biomass ratio, coarse root:branch biomass ratio and below-ground:above-ground biomass ratio. White circles represent the control group, and black circles represent the dry-treatment group. Red, green and blue lines are the fitting regressions for all trees, the control group and the dry-treatment group, respectively.

3.4. Simulation of Biomass and Ecosystem Services

Using the urban tree growth model *CityTree* [34], the above-ground, below-ground and overall biomasses for both the control and dry-treatment groups can be simulated. The measured and simulated biomasses of the trees, including the above-ground, below-ground and overall biomasses, had no significant differences, which showed the model's reliability in prediction ($p > 0.05$). Higher but not significantly different simulation was found in the dry-treatment group for the above-ground and overall biomasses (see Figure 7).

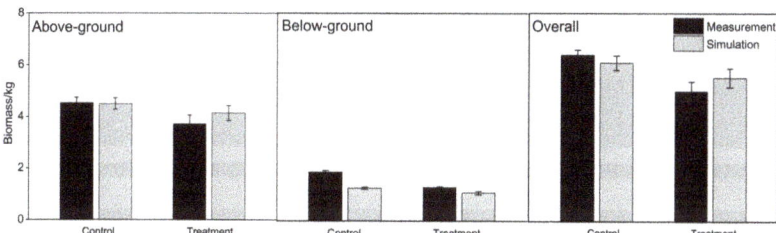

Figure 7. Comparison of measurements and simulations for the above-ground, below-ground and overall biomasses of trees from the control and dry-treatment groups.

The *CityTree* model also simulated biomass development for the entire year 2017 (Figure 8). In the control group, for the above- and below-ground biomasses, distinct increases of 0.6 kg tree^{-1} (from 3.9 to 4.5 kg tree^{-1}) and 0.4 kg tree^{-1} (from 0.9 to 1.3 kg tree^{-1}) were obtained, respectively. In the treatment group, however, trees showed scarce development, in that only a slight increase of 0.1 kg tree^{-1} was observed (from 5.4 to 5.5 kg tree^{-1}). For the root:shoot ratio, increases were found from 0.23 to 0.29 and from 0.02 to 0.27 for the control and treatment groups, respectively, from January to December.

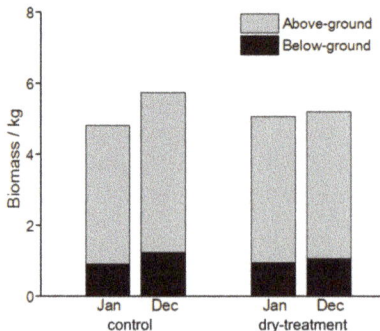

Figure 8. Simulated above- and below-ground biomasses for trees of the control and treatment groups for the entire year 2017.

By using the model *CityTree*, the ecosystem services of the trees, such as CO_2 fixation, water consumption and cooling potential by transpiration, could also be estimated. Figure 9 gives the values of these variables.

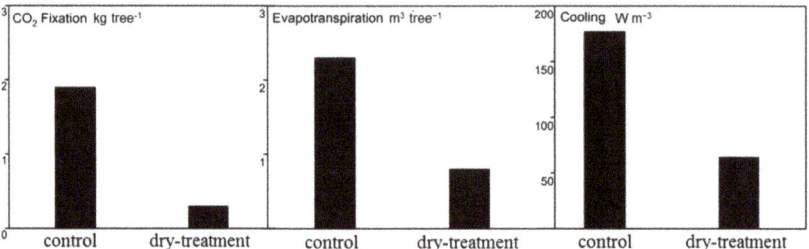

Figure 9. Ecosystem services, including CO_2 fixation, evapotranspiration and cooling potential by transpiration, for trees of the control and treatment groups in the year 2017.

The trees of the control group showed significantly higher CO_2 fixation than those of the dry-treatment group. Each tree in the control group could fix 1.9 kg CO_2 year^{-1}, which was far more than 0.3 kg CO_2 year^{-1} calculated for trees of the dry-treatment group. With a three times higher evapotranspiration (2.3 m^3 tree^{-1} compared with 0.8 m^3 tree^{-1}), the cooling potential provided by the control group trees was almost trebled compared with that of the dry-treatment group trees (177 W m^{-3} compared with 64 W m^{-3}). In summary, under the intense drought conditions associated with rainfall exclusion, all the ecosystem services of the lime trees were markedly reduced.

4. Discussion

4.1. Growth Patterns Under Drought

This study aimed to compare and analyse the growth patterns of *T. cordata* Greenspire from the control and dry-treatment groups. On the basis of the significant difference in soil water between the control and treatment groups ($p < 0.01$), our RER proved to be successful at simulating drought stress. The effects of drought manipulations are very complex [37], and the responses of plants to water scarcity are complicated, involving adaptive changes [38]. Under normal conditions, *T. cordata* Greenspire exhibited steady growth, especially in spring, which is consistent with previous research [18]. Since the RER was established, *T. cordata* Greenspire showed reduced stem growth in the dry-treatment group ($p < 0.05$). Furthermore, a large number of leaves died in the summer of 2017, in line with the decreased LAI. From the harvesting campaign, all the three compartments (branch, stem and coarse roots) showed significant reductions in their biomass in the dry-treatment group compared with the levels in the control group. Overall, the findings showed that *T. cordata* Greenspire did not cope well with a dry period and lost a great deal of biomass, which could be thought of as being associated with a negative water balance due to insufficient water supply. This is also in line with previous research showing that the water management of *T. cordata* Greenspire was disrupted following growth reduction [19,39].

The root:shoot ratio is often used to estimate relative biomass allocation between roots and shoots [37,40,41]. Previous studies have reported that, for some drought-resistant tree species, drought induced root growth to enhance water uptake, which led to an increase in the root:shoot ratio [42,43]. Mokany et al. [44] proposed that an increase in the root:shoot ratio could be a strategy when facing drought in the long term, whereas water shortage could shift carbon allocation to storage in the short term [45]. In our study, drought significantly ($p < 0.05$) reduced all three types of root:shoot ratio (fine root:leaf biomass ratio, coarse root:branch biomass ratio and below-ground:above-ground biomass ratio), which showed an opposite pattern to those drought-resistant tree species. Besides, the early discolouring and fall of leaves together with the biomass decrease in all the three compartments (i.e. coarse roots, stem and branches) suggested that *T. cordata* Greenspire did not have the means to positively adapt to drought. Additionally, the differences in water use efficiency between *T. cordata* Greenspire on the one hand and *Acer campestre* L. subsp. *campestre*, *O. carpinifolia* and *T. tomentosa* 'Brabant' from drier habitats on the other were measured and analysed [19]. The results showed that *T. cordata* Greenspire coped poorly with drought with highly reduced water use, whereas the other three tree species/cultivars maintained higher water use efficiency, possibly because of the better use of carbon for root production at the expense of above-ground biomass. Hence, the growth patterns of stem growth, biomass allocation and root:shoot ratios showed that *T. cordata* Greenspire failed to act as a drought-resistant cultivar.

4.2. Strategy Within the Root System

Fine roots are the main plant component involved in absorbing water and nutrients, whereas coarse roots provide stability [46,47]. Joslin et al. [48] and Germon et al. [49] proposed that the cost of fine root construction could be balanced by the uptake of water under drought stress. In this study, the control group showed gradual development from May to September and maintained its biomass from September to November. In contrast, fine root biomass was reduced from May to September but slightly increased from September to November in the dry-treatment group. Combined with the results of Figure 5, the root system in the dry-treatment group had fewer roots and nodes as well as decreased widths. Taking into account the decreased root:shoot ratio, it was shown that *T. cordata* Greenspire did not invest greatly in the root system.

Nevertheless, we found a particular pattern where the deep fine roots achieved substantial development, whereas the fine root biomass of the upper and middle layers clearly decreased. Gewin [50] showed that deep roots could be crucial to alleviating water stress in plants, especially for plants living in tropical and subtropical environments [51]. This suggests that, within the root system, *T. cordata* Greenspire prioritises deep fine roots to enhance its water uptake capacity rather than the shallow ones, which is in line with previous findings [52,53]. Furthermore, from September to November, it was observed that the deep roots had a decrease while the biomass from the upper and middle layer had an obvious development. This could be implied that the response of fine root to drought had a time difference among layers. Overall, despite inadequate investment in the root system, the dynamics of fine roots in different layers could reflect the positive behavior under drought.

4.3. Simulated Ecosystem Service Provision Under Drought Stress

Ecological process-based models have been applied widely for ecological issues, including biodiversity, phenology, hydrology and ecosystem services [54]. To obtain reliable predictions, however, the process-based models need to be validated first, in combination with observational data [55]. In our study, the *CityTree* model was first used to simulate the biomass from the below-ground and above-ground processes. The results showed high performance of the model, in that little difference was found between the observation and the simulation. Hence, the model should be reliable for simulating ecosystem services.

Figure 9. suggests that *T. cordata* Greenspire trees provided much less ecosystem services under heavy drought than unstressed trees in terms of CO_2 fixation, transpiration and cooling. This can be explained by the fact that, under conditions of water shortage, the lime trees had to expend more energy in seeking out water to survive. However, on the basis of clearly reduced total transpiration, the total biomass growth was markedly reduced, which led to a reduction in carbon fixation of 84% in the drought-stressed lime trees compared with the unstressed trees. Along with the smaller amounts of transpiration, the cooling effect was reduced by 64% for the drought-stressed trees. Besides, Figure 8 shows that the model predicted higher (albeit not significantly) biomass for the drought-stressed trees than for the control trees, which suggests that drought stress had a more severe impact in reality than predicted by the model. Taking this conservative prediction into account, the negative impact of drought stress on the ecosystem services provided by *T. cordata* Greenspire might be more serious.

5. Conclusions

In 2014, the IPCC (The Intergovernmental Panel on Climate Change) predicted that extreme climatic events such as severe drought would become more common and severe in the future. The urbanisation accompanying such global change can alter the composition, structure and biogeography of vegetation in cities and surrounding areas [56]. Hence, obtaining comprehensive knowledge about how urban trees react to and cope with dry conditions can be of utmost importance for ecosystem services in cities.

We analysed tree growth and simulated ecosystem services under drought conditions for *T. cordata* Greenspire. The cultivar appeared to be susceptible to drought, in that the biomasses of coarse roots, stem and branches decreased following a decrease in the root:shoot ratio and demonstrated substantial failure to provide ecosystem services. With the loss of much root biomass, it tended to invest in fine roots. At the beginning of drought, investment was given to the deep fine roots and in autumn the shallow fine roots obtained more development. Therefore, urban trees, particularly drought-susceptible tree species and cultivars such as *T. cordata* Greenspire, should be planted in large soil pits so that water shortages are minimised. Intensive maintenance of the trees based on their growth patterns as well as on the site and soil conditions could preserve tree vitality and enhance tree growth and the provision of ecosystem services.

Against a background of global climate change and increasing urbanisation, featuring an enhanced impact of urban heat on tree growth and vitality, *T. cordata* Greenspire will become a vulnerable urban tree species in Central European cities. Especially in temperate cities where currently the precipitation in summer is low and where future climate conditions may feature droughts of increasing number and intensity, tree species and cultivars such as *T. cordata* Greenspire will suffer severely. Long-term drought experiments are necessary to obtain detailed knowledge about the behaviour of urban tree species under intense drought conditions, and such information will be crucial for landscape planners and architects.

Author Contributions: Data curation, C.Z. and L.M.F.S.; Formal analysis, C.Z., L.M.F.S. and T.R.; Investigation, C.Z.; Methodology, C.Z., L.M.F.S. and T.R.; Project administration, T.R.; Resources, L.M.F.S. and T.R.; Software, T.R.; Supervision, H.P. and T.R.; Validation, C.Z.; Writing—original draft, C.Z.; Writing—review & editing, L.M.F.S. and T.R.

Acknowledgments: We thank the municipal nursery of Munich for the support and encouragement to conduct our field study there. The authors also express their gratitude to Yuan Ni for her help in creating the graphs and Jonas Schweiger for his assistance in field data collection.

Conflicts of Interest: The authors declare no conflicts of interest.

References

1. Change, I.C. *Impacts, Adaptation, and Vulnerability. Part A: Global and Sectoral Aspects. Contribution of Working Group II to the Fifth Assessment Report of the Intergovernmental Panel on Climate Change*; IPCC: Geneva, Switzerland, 2014.
2. Gregg, J.W.; Jones, C.G.; Dawson, T.E. Urbanization effects on tree growth in the vicinity of New York City. *Nature* **2003**, *424*, 183. [CrossRef] [PubMed]
3. Breshears, D.D.; Cobb, N.S.; Rich, P.M.; Price, K.P.; Allen, C.D.; Balice, R.G.; Romme, W.H.; Kastens, J.H.; Floyd, M.L.; Belnap, J. Regional vegetation die-off in response to global-change-type drought. *Proc. Natl. Acad. Sci. USA* **2005**, *102*, 15144–15148. [CrossRef] [PubMed]
4. Alberti, M. The effects of urban patterns on ecosystem function. *Int. Reg. Sci. Rev.* **2005**, *28*, 168–192. [CrossRef]
5. Greene, C.S.; Robinson, P.J.; Millward, A.A. Canopy of advantage: Who benefits most from city trees? *J. Environ. Manag.* **2018**, *208*, 24–35. [CrossRef] [PubMed]
6. Xing, Y.; Brimblecombe, P. Role of vegetation in deposition and dispersion of air pollution in urban parks. *Atmos. Environ.* **2019**, *201*, 73–83. [CrossRef]
7. Roy, S.; Byrne, J.; Pickering, C. A systematic quantitative review of urban tree benefits, costs, and assessment methods across cities in different climatic zones. *Urban For. Urban Green.* **2012**, *11*, 351–363. [CrossRef]
8. Tyrväinen, L.; Pauleit, S.; Seeland, K.; de Vries, S. Benefits and uses of urban forests and trees. In *Urban Forests and Trees*; Springer: Berlin/Heidelberg, Germany, 2005; pp. 81–114.
9. Dwyer, J.F.; Schroeder, H.W.; Gobster, P.H. The significance of urban trees and forests: Toward a deeper understanding of values. *J. Arboric.* **1991**, *17*, 276–284.
10. Nowak, D.J.; Greenfield, E.J.; Hoehn, R.E.; Lapoint, E. Carbon storage and sequestration by trees in urban and community areas of the United States. *Environ. Pollut.* **2013**, *178*, 229–236. [CrossRef]
11. Konarska, J.; Uddling, J.; Holmer, B.; Lutz, M.; Lindberg, F.; Pleijel, H.; Thorsson, S. Transpiration of urban trees and its cooling effect in a high latitude city. *Int. J. Biometeorol.* **2016**, *60*, 159–172. [CrossRef]
12. Velasco, E.; Roth, M.; Norford, L.; Molina, L.T. Does urban vegetation enhance carbon sequestration? *Landsc. Urban Plan.* **2016**, *148*, 99–107. [CrossRef]
13. Pretzsch, H.; Biber, P.; Uhl, E.; Dahlhausen, J.; Schütze, G.; Perkins, D.; Rötzer, T.; Caldentey, J.; Koike, T.; van Con, T. Climate change accelerates growth of urban trees in metropolises worldwide. *Sci. Rep.* **2017**, *7*, 15403. [CrossRef] [PubMed]
14. Zhao, M.; Running, S.W. Drought-induced reduction in global terrestrial net primary production from 2000 through 2009. *Science* **2010**, *329*, 940–943. [CrossRef] [PubMed]
15. Williams, A.P.; Allen, C.D.; Macalady, A.K.; Griffin, D.; Woodhouse, C.A.; Meko, D.M.; Swetnam, T.W.; Rauscher, S.A.; Seager, R.; Grissino-Mayer, H.D. Temperature as a potent driver of regional forest drought stress and tree mortality. *Nat. Clim. Chang.* **2013**, *3*, 292. [CrossRef]
16. Del Río, M.; Schütze, G.; Pretzsch, H. Temporal variation of competition and facilitation in mixed species forests in C entral E urope. *Plant Biol.* **2014**, *16*, 166–176. [CrossRef] [PubMed]
17. Rötzer, T.; Seifert, T.; Gayler, S.; Priesack, E.; Pretzsch, H. Effects of stress and defence allocation on tree growth: Simulation results at the individual and stand level. In *Growth and Defence in Plants*; Springer: Berlin/Heidelberg, Germany, 2012; pp. 401–432.
18. Moser, A.; Rötzer, T.; Pauleit, S.; Pretzsch, H. The urban environment can modify drought stress of small-leaved lime (*Tilia cordata* Mill.) and black locust (*Robinia pseudoacacia* L.). *Forests* **2016**, *7*, 71. [CrossRef]
19. Stratópoulos, L.M.F.; Duthweiler, S.; Häberle, K.-H.; Pauleit, S. Effect of native habitat on the cooling ability of six nursery-grown tree species and cultivars for future roadside plantings. *Urban For. Urban Green.* **2018**, *30*, 37–45. [CrossRef]
20. Johnson, A.D.; Gerhold, H.D. Carbon storage by urban tree cultivars, in roots and above-ground. *Urban For. Urban Green.* **2003**, *2*, 65–72. [CrossRef]
21. Barbaroux, C.; Bréda, N.; Dufrêne, E. Distribution of above-ground and below-ground carbohydrate reserves in adult trees of two contrasting broad-leaved species (*Quercus petraea* and *Fagus sylvatica*). *New Phytol.* **2003**, *157*, 605–615. [CrossRef]
22. West, J.B.; Hobbie, S.E.; Reich, P.B. Effects of plant species diversity, atmospheric [CO_2], and N addition on gross rates of inorganic N release from soil organic matter. *Glob. Chang. Biol.* **2006**, *12*, 1400–1408. [CrossRef]

23. Bardgett, R.D.; Wardle, D.A. *Aboveground-Belowground Linkages: Biotic Interactions, Ecosystem Processes, and Global Change*; Oxford University Press: Oxford, UK, 2010.
24. Eisenhauer, N.; Cesarz, S.; Koller, R.; Worm, K.; Reich, P.B. Global change belowground: impacts of elevated CO_2, nitrogen, and summer drought on soil food webs and biodiversity. *Glob. Chang. Biol.* **2012**, *18*, 435–447. [CrossRef]
25. Li, Z.; Kurz, W.A.; Apps, M.J.; Beukema, S.J. Belowground biomass dynamics in the Carbon Budget Model of the Canadian Forest Sector: recent improvements and implications for the estimation of NPP and NEP. *Can. J. For. Res.* **2003**, *33*, 126–136. [CrossRef]
26. Rowell, D.P.; Jones, R.G. Causes and uncertainty of future summer drying over Europe. *Clim. Dyn.* **2006**, *27*, 281–299. [CrossRef]
27. Meier, I.C.; Leuschner, C. Belowground drought response of European beech: fine root biomass and carbon partitioning in 14 mature stands across a precipitation gradient. *Glob. Chang. Biol.* **2008**, *14*, 2081–2095. [CrossRef]
28. Oliveira, S.; Andrade, H.; Vaz, T. The cooling effect of green spaces as a contribution to the mitigation of urban heat: A case study in Lisbon. *Build. Environ.* **2011**, *46*, 2186–2194. [CrossRef]
29. Rahman, M.A.; Moser, A.; Rötzer, T.; Pauleit, S. Microclimatic differences and their influence on transpirational cooling of Tilia cordata in two contrasting street canyons in Munich, Germany. *Agric. For. Meteorol.* **2017**, *232*, 443–456. [CrossRef]
30. Gill, S.; Rahman, M.; Handley, J.; Ennos, A. Modelling water stress to urban amenity grass in Manchester UK under climate change and its potential impacts in reducing urban cooling. *Urban For. Urban Green.* **2013**, *12*, 350–358. [CrossRef]
31. Gillner, S.; Vogt, J.; Tharang, A.; Dettmann, S.; Roloff, A. Role of street trees in mitigating effects of heat and drought at highly sealed urban sites. *Landsc. Urban Plan.* **2015**, *143*, 33–42. [CrossRef]
32. Poudel, K.; Temesgen, H. Methods for estimating aboveground biomass and its components for Douglas-fir and lodgepole pine trees. *Can. J. For. Res.* **2015**, *46*, 77–87. [CrossRef]
33. Rötzer, T.; Grote, R.; Pretzsch, H. The timing of bud burst and its effect on tree growth. *Int. J. Biometeorol.* **2004**, *48*, 109–118. [CrossRef]
34. Rötzer, T.; Rahman, M.; Moser-Reischl, A.; Pauleit, S.; Pretzsch, H. Process based simulation of tree growth and ecosystem services of urban trees under present and future climate conditions. *Sci. Total Environ.* **2019**, *676*, 651–664. [CrossRef]
35. Pretzsch, H.; Dieler, J.; Seifert, T.; Rötzer, T. Climate effects on productivity and resource-use efficiency of Norway spruce (*Picea abies* [L.] Karst.) and European beech (*Fagus sylvatica* [L.]) in stands with different spatial mixing patterns. *Trees* **2012**, *26*, 1343–1360. [CrossRef]
36. Team, R.C. *R: A Language and Environment for Statistical Computing*; R Foundation for Statistical Computing: Vienna, Austria, 2013.
37. Poorter, H.; Niklas, K.J.; Reich, P.B.; Oleksyn, J.; Poot, P.; Mommer, L. Biomass allocation to leaves, stems and roots: Meta-analyses of interspecific variation and environmental control. *New Phytol.* **2012**, *193*, 30–50. [CrossRef] [PubMed]
38. Chaves, M.M.; Pereira, J.S.; Maroco, J.; Rodrigues, M.L.; Ricardo, C.P.P.; Osório, M.L.; Carvalho, I.; Faria, T.; Pinheiro, C. How plants cope with water stress in the field? Photosynthesis and growth. *Ann. Bot.* **2002**, *89*, 907–916. [CrossRef] [PubMed]
39. Moser, A.; Rahman, M.A.; Pretzsch, H.; Pauleit, S.; Rötzer, T. Inter-and intraannual growth patterns of urban small-leaved lime (*Tilia cordata* mill.) at two public squares with contrasting microclimatic conditions. *Int. J. Biometeorol.* **2017**, *61*, 1095–1107. [CrossRef] [PubMed]
40. Wilson, J.B. A review of evidence on the control of shoot: root ratio, in relation to models. *Ann. Bot.* **1988**, *61*, 433–449. [CrossRef]
41. Gowda, V.R.; Henry, A.; Yamauchi, A.; Shashidhar, H.; Serraj, R. Root biology and genetic improvement for drought avoidance in rice. *Field Crops Res.* **2011**, *122*, 1–13. [CrossRef]
42. Asch, F.; Dingkuhn, M.; Sow, A.; Audebert, A. Drought-induced changes in rooting patterns and assimilate partitioning between root and shoot in upland rice. *Field Crops Res.* **2005**, *93*, 223–236. [CrossRef]
43. Lemoine, R.; La Camera, S.; Atanassova, R.; Dédaldéchamp, F.; Allario, T.; Pourtau, N.; Bonnemain, J.-L.; Laloi, M.; Coutos-Thévenot, P.; Maurousset, L. Source-to-sink transport of sugar and regulation by environmental factors. *Front. Plant Sci.* **2013**, *4*, 272. [CrossRef]

44. Mokany, K.; Raison, R.J.; Prokushkin, A.S. Critical analysis of root: shoot ratios in terrestrial biomes. *Glob. Chang. Biol.* **2006**, *12*, 84–96. [CrossRef]
45. McDowell, N.; Pockman, W.T.; Allen, C.D.; Breshears, D.D.; Cobb, N.; Kolb, T.; Plaut, J.; Sperry, J.; West, A.; Williams, D.G. Mechanisms of plant survival and mortality during drought: why do some plants survive while others succumb to drought? *New Phytol.* **2008**, *178*, 719–739. [CrossRef]
46. Jackson, R.B.; Mooney, H.; Schulze, E.-D. A global budget for fine root biomass, surface area, and nutrient contents. *Proc. Natl. Acad. Sci. USA* **1997**, *94*, 7362–7366. [CrossRef] [PubMed]
47. Reubens, B.; Poesen, J.; Danjon, F.; Geudens, G.; Muys, B. The role of fine and coarse roots in shallow slope stability and soil erosion control with a focus on root system architecture: A review. *Trees* **2007**, *21*, 385–402. [CrossRef]
48. Joslin, J.; Gaudinski, J.B.; Torn, M.S.; Riley, W.; Hanson, P.J. Fine-root turnover patterns and their relationship to root diameter and soil depth in a 14C-labeled hardwood forest. *New Phytol.* **2006**, *172*, 523–535. [CrossRef] [PubMed]
49. Germon, A.; Cardinael, R.; Prieto, I.; Mao, Z.; Kim, J.; Stokes, A.; Dupraz, C.; Laclau, J.-P.; Jourdan, C. Unexpected phenology and lifespan of shallow and deep fine roots of walnut trees grown in a silvoarable Mediterranean agroforestry system. *Plant Soil* **2016**, *401*, 409–426. [CrossRef]
50. Gewin, V. Food: An underground revolution. *Nat. News* **2010**, *466*, 552–553. [CrossRef] [PubMed]
51. Pierret, A.; Maeght, J.-L.; Clément, C.; Montoroi, J.-P.; Hartmann, C.; Gonkhamdee, S. Understanding deep roots and their functions in ecosystems: an advocacy for more unconventional research. *Ann. Bot.* **2016**, *118*, 621–635. [CrossRef]
52. Goisser, M.; Geppert, U.; Rötzer, T.; Paya, A.; Huber, A.; Kerner, R.; Bauerle, T.; Pretzsch, H.; Pritsch, K.; Häberle, K. Does belowground interaction with *Fagus sylvatica* increase drought susceptibility of photosynthesis and stem growth in *Picea abies*? *For. Ecol. Manag.* **2016**, *375*, 268–278. [CrossRef]
53. Leuschner, C.; Hertel, D.; Schmid, I.; Koch, O.; Muhs, A.; Hölscher, D. Stand fine root biomass and fine root morphology in old-growth beech forests as a function of precipitation and soil fertility. *Plant Soil* **2004**, *258*, 43–56. [CrossRef]
54. He, Y.; Yang, J.; Zhuang, Q.; McGuire, A.D.; Zhu, Q.; Liu, Y.; Teskey, R.O. Uncertainty in the fate of soil organic carbon: A comparison of three conceptually different decomposition models at a larch plantation. *J. Geophys. Res. Biogeosci.* **2014**, *119*, 1892–1905. [CrossRef]
55. Parmesan, C.; Burrows, M.T.; Duarte, C.M.; Poloczanska, E.S.; Richardson, A.J.; Schoeman, D.S.; Singer, M.C. Beyond climate change attribution in conservation and ecological research. *Ecol. Lett.* **2013**, *16*, 58–71. [CrossRef]
56. Allen, C.D.; Macalady, A.K.; Chenchouni, H.; Bachelet, D.; McDowell, N.; Vennetier, M.; Kitzberger, T.; Rigling, A.; Breshears, D.D.; Hogg, E.T. A global overview of drought and heat-induced tree mortality reveals emerging climate change risks for forests. *For. Ecol. Manag.* **2010**, *259*, 660–684. [CrossRef]

© 2019 by the authors. Licensee MDPI, Basel, Switzerland. This article is an open access article distributed under the terms and conditions of the Creative Commons Attribution (CC BY) license (http://creativecommons.org/licenses/by/4.0/).

Article

Structure, Diversity, and Carbon Stocks of the Tree Community of Kumasi, Ghana

Bertrand Festus Nero [1,*], Daniel Callo-Concha [2] and Manfred Denich [2]

1. Department of Land Reclamation and Rehabilitation, Kwame Nkrumah University of Science and Technology, PMB, University Post Office, Kumasi, Ghana
2. Center for Development Research (ZEF), University of Bonn, Genscherallee 3, D-53113 Bonn, Germany; d.callo-concha@uni-bonn.de (D.C.-C.); uzef0008@uni-bonn.de (M.D.)
* Correspondence: bfnero@knust.edu.gh or bfn8puo@gmail.com; Tel.: +233-50-395-8016

Received: 23 June 2018; Accepted: 8 August 2018; Published: 29 August 2018

Abstract: Urban forestry has the potential to address many urban environmental and sustainability challenges. Yet in Africa, urban forest characterization and its potential to contribute to human wellbeing are often neglected or restrained. This paper describes the structure, diversity, and composition of an urban forest and its potential to store carbon as a means of climate change mitigation and adaptation in Kumasi. The vegetation inventory included a survey of 470,100-m² plots based on a stratified random sampling technique and six streets ranging from 50 m to 1 km. A total of 3757 trees, comprising 176 species and 46 families, were enumerated. Tree abundance and species richness were left skewed and unimodally distributed based on diameter at breast height (DBH). Trees in the diameter classes >60 cm together had the lowest species richness (17%) and abundance (9%), yet contributed more than 50% of the total carbon stored in trees within the city. Overall, about 1.2 million tonnes of carbon is captured in aboveground components of trees in Kumasi, with a mean of 228 t C ha^{-1}. Tree density, DBH, height, basal area, aboveground carbon storage, and species richness were significantly different among green spaces ($p < 0.05$). The diversity was also significantly different among urban zones ($p < 0.0005$). The DBH distribution of trees followed a modified reverse J-shaped model. The urban forest structure and composition is quite unique. The practice of urban forestry has the potential to conserve biological diversity and combat climate change. The introduction of policies and actions to support the expansion of urban forest cover and diversity is widely encouraged.

Keywords: species richness; abundance; urban forest; green spaces; sustainability

1. Introduction

More than 50% of Ghana's population now reside in urban areas [1,2]. Kumasi, the second largest city in Ghana, is expanding rapidly in both land area and population. Its population and land area have respectively expanded from about 300,000 inhabitants and 25 km² in the 1950s to about 2.5 million and 254 km² today. Between 1986 and 2014, about 200 ha of Kumasi's green cover was lost annually to urban build-up and road infrastructure [3]. In spite of this massive loss, some 33% of the land area of the political metropolitan Kumasi is still predominantly vegetated, of which 65% is composed of woody tree/shrub cover [3,4]. This remaining natural land cover is a vital ingredient for environmental sustainability and a primary source of key ecosystem goods and services. Yet knowledge about the forest structure and composition, as well as its carbon storage potential, is limited.

Successful ecological management of the urban forest requires a thorough understanding of the structure and composition of remnant trees, forest patches, and the dynamic variability between and within green space types. This reveals inter-green space structural and compositional differences and provides information on how to achieve conservation goals. The urban forest structure and composition is considerably unique and distinct from that of natural and plantation forests. Unlike the latter two, the urban forest exists as fragments of tree clusters of varying sizes scattered between grey infrastructures and may assume several shapes of varying dimensions [5]. The tree community is composed of planted and naturally regenerated species of both exotic and native origins, hence urban forest can be more diverse than forest in neighboring open landscape [6]. Species composition and diversity depend on the ecological zone. For instance, in West Africa, the species tolerant to droughts predominate in cities of the tropical dry savannah, while cities in the tropical high forest zone are composed of species adapted to moist and humid conditions [7]. The size-class distribution of the urban forest is a blend of plantation and natural forests and a mixture of small and large growing trees [6]. The extent, species composition, and structure are influenced by the city morphology, natural environmental characteristics, human management and city age [8], and the pre-development land use types [9]. The stratified levels of stress among green spaces [10] and the lack of or ineffective implementation of urban forestry policies in developing countries [7] affect the pattern of urban forest structure, diversity, and species composition.

Assessing urban forest structure and composition involves measuring and recording information on every tree and the methods adopted are quite well-documented. A complete census of all trees is the most precise way of collecting data, but is cost- and time-consuming for larger tree populations. Random sampling techniques have been used by several studies in the past to assess urban forest structure and composition [11–15]. Other recent studies have used models such as the urban forest effects model, designed and used in several cities around the world, to evaluate urban forest structure, composition, function, and values [13,14]. Remote sensing techniques have also been used intensively to map and depict the distribution, structure, and function of urban forest at a citywide scale [16–19]. These studies depict the urban forest as a spatially- and vertically-stratified multifunctional entity composed of patches of different sizes, species compositions, tree densities, and tree sizes. Despite this progress, there is limited knowledge on the forest structure, species composition, and forest functions in urban areas in developing countries and urban and peri-urban forest resources are neglected in national forest policies and strategies in countries such as Ghana [20].

There have been several vegetation inventory studies conducted in the forest landscape of Ghana. For instance, between 1985 and 1989, the forest inventory project was implemented to provide data for sustained yield policy formulation and to establish inventory units [21]. This was continued under a new name, as the forest inventory and management project (1989–1997) [22]. The latter provided a more comprehensive national level estimate of the forest structure, species composition, and the yield potential of timber resources in forest inside and outside forest reserves. However, none of the past inventory programs have accounted for urban landscapes in the country, despite the potential of the urban forest to enhance urban sustainability and resilience. The present study contributes to this gap by assessing the tree species composition, diversity, and structure of the urban forest of Kumasi. More specifically, the study (1) describes the tree community structure, species diversity, and composition differences among green spaces of Kumasi; (2) establishes the linkage between urban forest structure, species diversity, and carbon storage (productivity); and (3) discusses the conservation relevance of the urban forest structure and composition of Kumasi.

2. Materials and Methods

2.1. Study Area

Kumasi metropolis is located in south central Ghana (6°41″ N, 1°37″ W, Figure 1). The climate is tropical, characterized by a bi-modal rainfall system. Mean annual rainfall and temperature are 1250 mm [23] and 26.4 °C [24], respectively. Kumasi's current population accounts for 10% of the total population of Ghana. It is the fastest growing and second largest city in the country, with a population density of 10,000 persons per km^2 and an annual growth rate of 4.8% [1,25]. It is a central transiting point for travelers from within and beyond Ghana and hosts the largest open market in West Africa [26].

Kumasi's history as the "Garden City of West Africa", in addition to its rapid population growth rate, medium-size character, and location in the tropical high forest zone, make it an ideal city for the analysis of urban forest structure and composition. The metropolitan land area of Kumasi is about 254 km² (Figure 1) and is partitioned into ten sub-metropolises.

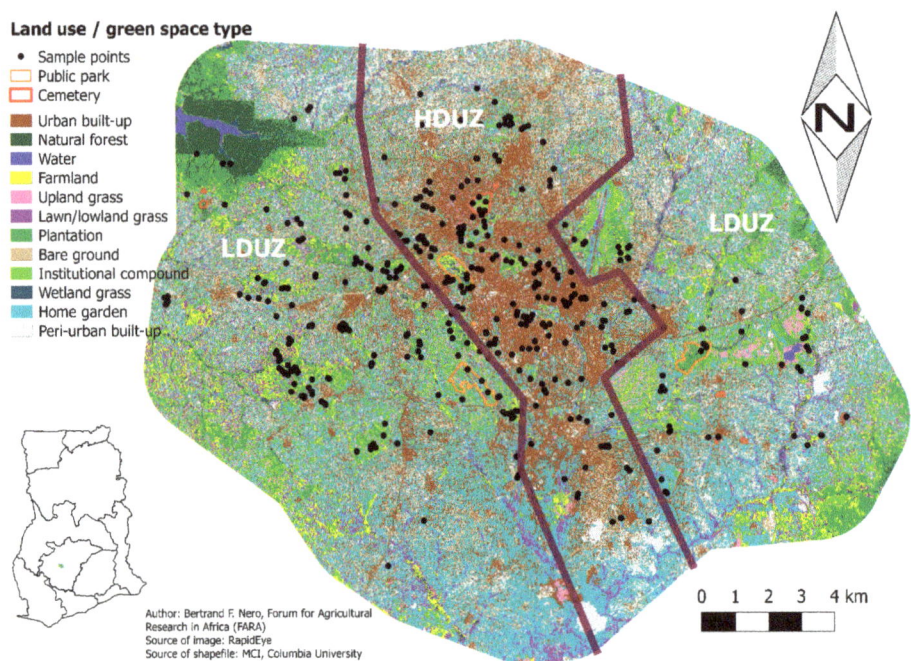

Figure 1. Urban green space map of Kumasi showing the two urban zones. HDUZ—high density urban zone and LDUZ—low density urban zone.

2.2. Vegetation Sampling

Plots were located in Kumasi by a stratified random sampling design (Figure 1) in 2014 as part of the urban biomass study in Ghana [4]. The city was partitioned into two zones based on the normalized difference vegetation index (NDVI): High Density Urban Zone (HDUZ or core urban; mean NDVI ≤ 0.11) and Low Density Urban Zone (LDUZ or peri-urban; mean NDVI > 0.11) (Figure 1). About 12 land-use classes (including eight urban green space (UGS) types) were delineated using satellite imagery techniques [3]. The UGS types included: plantations, natural forest, home gardens, institutional compounds, farmlands, cemeteries/sacred groves, public parks, and grasslands/rangelands (Figure 1). Forest (plantation or natural) refers to any extensive area of planted or naturally occurring trees, either managed or unmanaged and occupying an area of at least 0.5 ha with trees at least 5 m tall and a canopy cover of at least 80%.

A total of 470 sample points were randomly generated on the UGS map of Kumasi, Ghana, for the assessment of forest structure and biodiversity (Figure 1). A 10 × 10 m quadrat was established for each sampling point with the help of a compass, a distance tape measure, and ranging poles, except in home gardens, where the entire area of the garden was surveyed. All trees with a diameter at breast height (DBH = 1.3 m from ground) >5 cm within each plot were identified to the species level, counted by species, and the height and DBH were measured. For trees with multiple stems below the DBH mark, the diameter of each stem was measured and their combined mean estimated as the quadratic mean of the individual stem diameters. At least six streets of lengths ranging from 50 m to 1 km were purposively selected and all trees along each road were counted by species and their heights and DBH measured. Purposive sampling was required here because the majority of streets in Kumasi are devoid of trees. The species identification was carried out with the aid of tree experts and published tree identification guides, such as those by Hawthorne and Gyakari [27] and Oteng-Amoako [28].

2.3. Data Analysis

Tree density, basal area, and size class distribution (SCD) were estimated and constituted the basis for describing the urban forest structure of Kumasi. Tree density is the number of trees per plot area, while basal area is expressed as $0.00007854 \times D^2$ (D = DBH in cm). An interval of 10 cm was maintained between successive diameter size classes, except for diameters above 90 cm, where larger intervals >10 cm were adopted. Species richness, abundance, and biomass in each DBH class were calculated for the entire city and for each green space type.

The aboveground biomass of each tree was computed using the generalized biomass model developed for pantropical forest trees (Equation (1)) [29].

$$AGB = 0.0673 \times (\rho D^2 h)^{0.976} \tag{1}$$

where AGB = aboveground biomass (kg), D = DBH in cm, h = height in m, and ρ = dry wood density of the tree species. For plants in the Arecaceae family, height was measured *and* specific biomass equations involving height used for the estimation of biomass, since the DBH of these trees are usually inaccessible and unmeasurable.

The dry wood density of each species was obtained from published literature and global databases [30,31]. The biomass was then multiplied by 0.474 to obtain carbon stocks per tree [32].

Species richness was expressed as the number of observed species for each DBH class of the sampled trees, while *Chao1* was used to estimate the potential species richness for the entire city, each UGS type, and each urban zone (Equation (2)). *Chao1*, the simplest nonparametric estimator, estimates the total number of species (Sest) by adding a term that only depends on the observed number of singletons (a), species each represented by a single individual) and doubletons (b), species each represented by exactly two individuals) to the number of species observed (Sobs) [33].

$$Sest = Sobs + \frac{a^2}{2b} \tag{2}$$

Species accumulation curves were constructed based on the number of individuals and the sample size/number. Shannon's (H) and Simpson's (D) diversity indices and Pielou's (J) evenness were calculated. Chi-square statistics were used to establish significant differences in species richness and diversity among green spaces and urban zones.

Beta diversity analysis was performed to determine similarities in species composition among urban zones and UGS types. Beta diversity was estimated with the reformulated Sørenson's and Jaccard's indices instead of the binary techniques often employed [34]. These indices use a probabilistic approach which combines incidence-based indices with relative abundance data, thus minimizing bias and placing unequal weightings on rare and common species [34]. The computations of these indices are outlined in Chao et al. [34]. The values of both indices range between 0 and 1: with a value of 0 implying absolute dissimilarity and a value of 1 implying absolute similarity [33,35]. Thus, high values reflect a low beta diversity (high similarity) and low values reflect a high beta diversity (high dissimilarity). Bootstrap confidence intervals (CI) of the indices at 95% CI were estimated using 1000 iterations.

Correspondence analysis (CA) was performed to show the association between tree species, UGS type, and urban zone. CA graphically displays the relationship between variables which otherwise would not be detected using a pairwise test of associations. The graphs represent relative frequencies based on the distance between row (green space or urban zone) and column (species) profiles and the

3. Results

3.1. Species Diversity and Composition

Overall, 176 species in 46 families were enumerated in Kumasi. Species richness differed significantly among green space types and urban zones. The species accumulation curves for the different green space types are presented (Figure 2), indicating the increment in the number of species with sampling effort. The study further indicates that sampling was adequately done. Home gardens, public parks, and institutional compounds were the most species rich; 80, 79, and 75, respectively (Figure 2, Table 1). The natural forest, however, had the highest species richness, as well as the highest Simpson's and Shannon's diversity indices (Table 1).

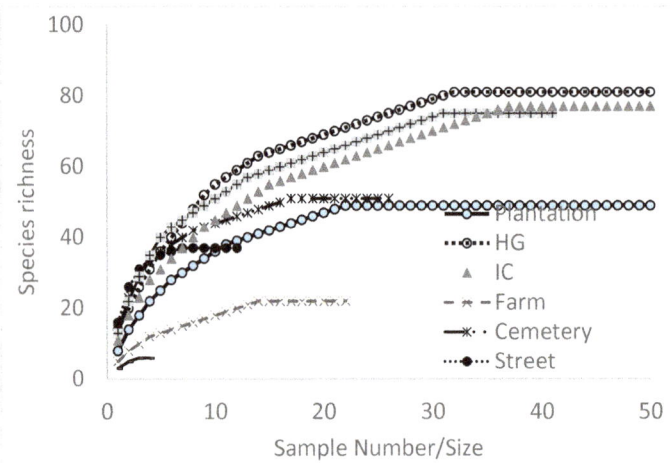

Figure 2. Rarefaction curves of woody species among green spaces in Kumasi. IC = Institutional compound, HG = home garden.

Table 1. Tree species abundance, richness, and diversity indices in different green space types within Kumasi. Chi-square analysis of richness indicates significant differences ($p < 0.0001$, $n = 8$, $X^2 = 139.4$).

Greenspace Type	Number of Individuals	Observed Richness, S	Estimated Chao1, Sest	Shannon H	Simpson 1-D	Pielou J (Evenness)
Plantation	630	48	73.6	2.561	0.854	0.66
Natural forest	980	96	105	3.840	0.969	0.84
Home garden	1095	80	98.6	3.158	0.919	0.72
Institutional Compound	715	79	101.3	3.502	0.951	0.80
Farm	100	23	47.0	2.269	0.821	0.72
Cemetery	266	51	81.3	3.242	0.935	0.82
Street	565	37	57.2	2.809	0.903	0.78
Public park	334	75	127.7	3.521	0.952	0.82
Grassland	39	6	8.3	0.749	0.328	0.42
Total	3757	176	222.4	3.716	0.956	0.72

Table 2 provides a list of the five most common species in each of the green space types. Mostly edible tree species are found in the home gardens, whereas species of aesthetic and shade values predominate along streets and within institutional compounds. Correspondence analysis reveals similarity among several green space types (Figure 3). About 51% of the association was well-represented in two dimensions. Dimension 1 (x-axis) representing the UGS type explained 28.5% of the total variation. Cemetery and natural forest were more similar in species composition courtesy of the predominance of native species, while home gardens (HG) and farmlands are also more similar in species composition due to the dominance of fruit and other agroforestry tree species. Common ornamentals and readily establishing exotic species result in species composition similarity among streets, plantations, institutional compounds (IC), and public parks (Figure 3, Table 2). Dimension 2 (y-axis), representing species, accounted for 22.1% of the total variation. The rule of thumb for the interpretation of the biplot is that species near each other are the most similar, UGS near each other are also the most similar, and species near a particular UGS type are most closely associated with or occur in that UGS type.

Table 2. Top five most common (abundant) species in each green space type in Kumasi.

UGS	Species
Home garden	Oil palm—*Elaeis guineensis* Jacq., Mango—*Mangifera indica* L., Pear—*Persea americana* Mill., Orange—*Citrus sinensis* (L) Osbeck, Coconut—*Cocos nucifera* L.
Street	Cassia—*Cassia siamea* Lam., Copper pod tree—*Peltophorum pterocarpum* (DC.) Backer ex K. Heyne, Cedrela—*Cedrela odorata* L., Indian almond—*Terminalia catappa* L., Flamboyant- *Delonix regia* (Bojer ex Hook).
Plantation	Teak—*Tectona grandis* L. f., Cocoa—*Theobroma cacao* L., *Cassia siamea*, *Elaeis guineensis*, White teak—*Gmelina arborea* (Roxb)
Institutional compound	*Cassia siamea*, *Millettia thonningii* (Schumach.) Baker, Weeping willow—*Polyalthia longifolia* (Sonn.) Thwaites, *Elaeis guineensis*, *Casuarina equisetifolia* L.
Cemetery	*Elaeis guineensis*, Brimstone tree—*Morinda lucida* (Benth.), *Cassia siamea*, *Mangifera indica*, Quickstick-*Gliricidia sepium* (Jacq.) Kunth ex Walp.
Public park	*Cassia siamea*, *Elaeis guineensis*, *Millettia thonningii*, *Pelthophorum pterocarpum*, *Delonix regia*
Farmland	*Elaeis guineensis*, *Morinda lucida*, *Mangifera indica*, African tulip—*Spathodea campanulata* P. Beauv, Kapok tree—*Ceiba pentandra* (L.) Gaertn.
Grassland	*Ceiba pentandra*, *Pithecellobium dulce* (Roxb.) Benth., *Cedrela odorata*, *Mangifera indica*, *Morinda lucida*
Natural forest	*Cedrela odorata*, *Elaeis guineensis*, Wawa—*Triplochiton scleroxylon* K. Schum., *Funtumia elastic* (Preuss) Stapf, *Trichilia monodelpha* (Thonn.) J. de Wilde

3.2. Urban Forest Structure and Diversity

The estimated total tree population in the study area is about $3,564,277 \pm 27,888$, with a mean of 377 trees ha^{-1}. The number of trees per unit area differed significantly among the green space types (Table 3, $p < 0.0001$). Plantations had the highest tree density of 825 trees ha^{-1}, followed by public parks with 383 trees ha^{-1} and by institutional compounds with 321 trees ha^{-1}. Farmlands had the lowest tree density of 175 trees ha^{-1}.

Basal area and DBH were also significantly different ($p > 0.0001$) among green spaces and urban zones. The mean DBH and basal area of the studied plots were 33.3 cm and 55.5 m^2 ha^{-1}, respectively. Public parks, natural forest, cemeteries, and institutional compounds had the largest stand basal areas, while grassland, home gardens, and farmland had the smallest basal areas. The natural forest had the largest sized trees, with a mean DBH of 47.5 cm (Table 3).

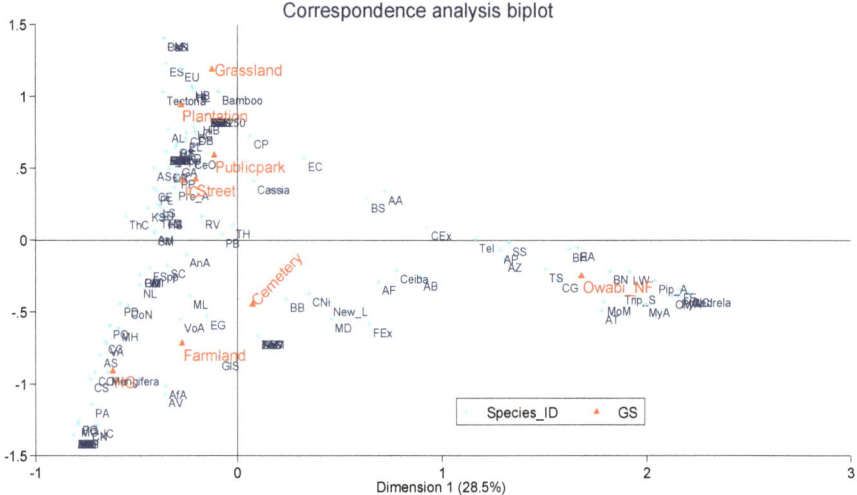

Figure 3. Output of correspondence analysis for the tree species in different green spaces (GS) in Kumasi metropolis. Dimension 1 represents green spaces while dimension 2 represents plant species. Chi-square = 11,169.8, Degrees of freedom = 2624. **Species names include**: AL, *Albizia lebbeck*; ASe, *Acasia senegale*; ASo, *Acalypha sonderina*; AD, *Adansonia digita*; AdP, *Adenanthera pavonina*; AfA, *Afzelia africana*; AF, *Albizia ferruginea*; AA, *A. adianthifolia*; AZ, *A. zygia*; AiC, *Alchornea cordifolia*; AlP, *Allanblackia parviflora*; AB, *Alstonia boonei*; AP, *Amphimas pterocarpoides*; AO, *Anacardium occidentale*; AM, *Annona muricata*; AS, *A. squamosal*; AnA, *Antiaris africana*; AT, *A. toxicaria*; AC, *Araucaria columnaris*; AI, *Artocarpus incisis*; AzI *Azadirachta indica*; AN, *Anthocleista vogelii*; AV, *A. nobilis*; Bamboo, *Bambusa vulgaris*; BN, *Baphia nitida*; BaS, *Baphia* spp.; BT, *Bauhinia tomentosa*; BS, *Blighia sapida*; BB, *Bombax buonopozense*; CaC, *Callitris cupressiformis*; CaP, *Calotropis procera*; CO, *Cananga odorata*; CP, *Carapa procera*; CaN, *Cassia nodiflora*; *C. siamea*; CE, *Casuarina equisetifolia*; CeP, *Cecropia peltata*; CeO, *Cedrela odorata*; Ceiba, *Ceiba pentandra*; CM, *Celtis mildbraedii*; CEx, *Chlorophora excelsa*; CZ, *Cinnamomum zeylanicum*; CL, *Citrus limon*; CN, *C. nobilis*; CS, *C. sinensis*; ClP, *Cleistopholis patens*; CF, *Cnestis ferruginea*; CoN, *Cocos nucifera*; CV, *Codiaeum variegatum*; CG, *Cola gigantea*; CoM, *C. millenii*; CNi, *C. nitida*; CA, *C. acuminata*; CMi, *Cordia millenii*; Cot, *Gossypium* spp; CC, *Crescentea cujete*; DO, *Daniella ogea*; DR, *Delonix regia*; DE, *Duranta erecta*; DG, *Dialium guineense*; DB, *Distemonanthus benthamianus*; DM, *Duboscia macrocarpa*; EG, *Elaeis guineensis*; FEx, *Ficus exasperata*; FSpp, *Ficus* spp.; FU, *F. umbellata*; FE, *Funtumia elastica*; GM, *Garcinia mangostana*; GlS, *Gliricidia sepium*; GA, *Gmelina arborea*; HL, *Hallea ledermannii*; HS, *H. stipulosa*; HB, *Hevea brasiliensis*; HiB, *Hildegardia barteri*; HF, *Holarrhena floribunda*; HyA, *Hymenostegia afzelii*; HA, *Hymenostegia aubrevillei*; JC, *Jatropha curcas*; KC, *Khaya cordifolia*; KS, *K. senegalensis*; LS, *Lagerstroemia speciose*; LaS, *Lannea schimperi*; LW, *L. welwitschii*; L_Spp, *Livingstonia* spp; MB, *Macaranga barteri*; MH, *M. heudelotii*; ME, *Maesopsis eminii*; MA, *Mammea africana*; *Mangifera indica*; MD, *Margaritaria discoidea*; MC, *Michelia champaca*; MT, *Millettia thonningii*; MH, *Millingtonia hortensis*; MM, *Monodora myristica*; ML, *Morinda lucida*; MO, *Moringa oleifera*; MoM, *Morus mesozygia*; MyA, *Myrianthus arboreus*; NL, *Nauclea latifolia*; New_L, *Newbouldia lavis*; OS, *Oncoba spinosa*; PB, *Parkia biglobosa*; PaS, *Parkinsonia speciosa*; PP, *Pelthophorum pterocarpum*; PA, *Persea americana*; PC, *Pinus caribaea*; Pip_A, *Piptadeniastrum africanum*; PD, *Pithecellobium dulce*; PS, *P. saman*; PlA, *Plumera alba*; PL, *Polyalthia longifolia*; PO, *P. oliveri*; Pro_A, *Prosopis africana*; PM, *Pseudospondias mombin*; PG, *Psidium guajava*; PsS, *Psydrax subcordata*; PH, *Pteleopsis hylodendron*; PyA, *Pycnanthus angolensis*; RV, *Rauwolfia vomitoria*; RL, *Rothmannia longiflora*; SD, *Samanea dinklagei*; SE, *Solanum erianthum*; SC, *Spathodea campanulata*; SM, *Spondias mombin*; SS, *Sterculia* spp.; TI, *Tamarindus indica*; Tectona, *Tectona grandis*; TeI, *Terminalia ivorensis*; TM, *T. montalis*; TeC, *T. catappa*; TS, *T. superba*; TA, *T. angolensis*; TT, *Tetrapleura tetraptera*; ThC, *Theobroma cacao*; TO, *Thuja orientalis*; TH, *Trichilia heudelotii*; Trip_S, *Triplochiton scleroxylon*; VA, *Vernonia amygdalina*; VoA, *Voacanga africana*. **GS type**; Owabi_NF = Natural forest, HG = Home garden, IC = Institutional compound.

Table 3. The structural attributes of the urban forest in Kumasi. Numbers in the same column followed by the same letter are not significantly different at $p = 0.05$.

UGS	Tree Density No. ha^{-1}	DBH cm	Height m	BA m^2 ha^{-1}	H:DBH m/cm
Streets		44.1ab	15.2bc		0.34
Cemeteries	261cd	44.0ab	13.2cde	83.8a	0.30
Farmlands	175d	45.5ab	11.4de	38.6c	0.25
Grasslands	200d	33.0cd	12.9cde	23.4c	0.39
Home gardens	240d	27.1cd	10.0e	24.0c	0.37
Institutional compounds	321c	35.3bc	13.6cd	71.3ab	0.39
Natural forest	246d	47.5a	44.5a	83.5a	0.94
Plantations	825a	23.2d	14.4bcd	65.0b	0.62
Public parks	484b	35.6bc	17.1b	84.9a	0.48
Mean	377	33.3	13.0	55.6	0.39
Least Square Difference		10.6	3.5		

Analysis of the diameter-size class distribution of the urban forest of Kumasi revealed a modified reverse J-shaped curve (Figure 4). Trees with extremely small girth sizes (DBH < 10 cm) were few, since the urban forest in many cases lacks understory vegetation. However, the medium diameter class (DBH > 10 cm) peaked at the DBH = 20 cm with over 400 trees and declined progressively till DBH = 50 cm, where the number of trees per DBH class appears to stabilize. Among green space types, some, such as plantations, institutional compounds, and streets, depicted a similar modified reverse J-model in their diameter distributions, while in others, such as cemeteries and home gardens, the reverse J-shaped model was clearly depicted (Figure 5). This suggests that the reverse J-shaped model does not describe the structure of all forest types. The urban landscape is a complex mixture of planted and conserved self-established trees of native species origin. Besides, due to the different levels of exposure to stress (e.g., pollution), it is unlikely that there is equal mortality among diameter classes.

Similarly, both the species richness and Shannon diversity index were left skewed, with peaks at DBH = 25 and 30 cm, respectively. Species richness ranged between 60 species in DBH = 25 cm and 19 species in DBH > 130 cm (Figure 4). Likewise, the Shannon index was 3.47 in DBH < 25 cm and 2.45 in the DBH > 130 cm.

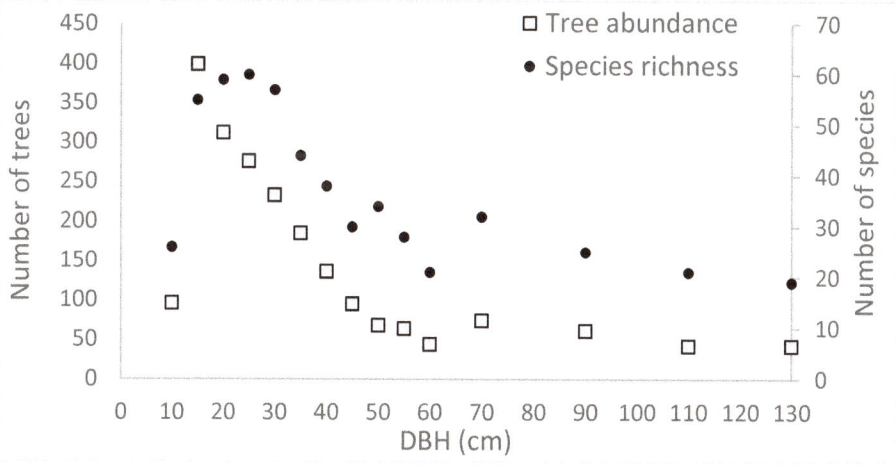

Figure 4. A unimodal distribution of the number of trees and number of species within Kumasi, Ghana.

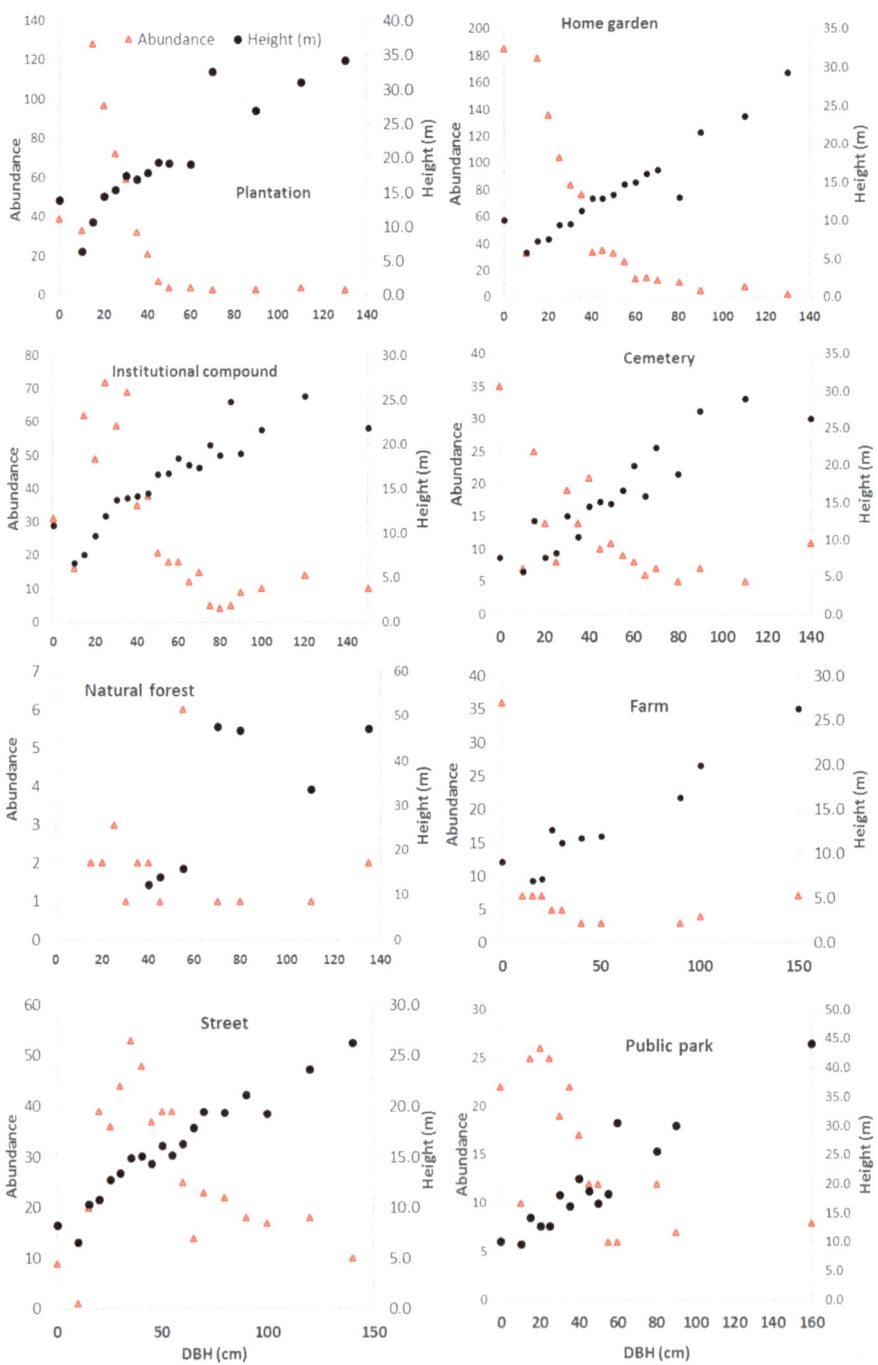

Figure 5. The abundance and mean total height of trees in different diameter at breast height (DBH) classes among green spaces of Kumasi, Ghana.

3.3. The Structure and Composition of the Urban Zones

The species richness and diversity of the HDUZ, LDUZ, and natural forest were significantly different. Species richness and diversity of the HDUZ was less than the LDUZ ($p < 0.0001$, $n = 1$, $X^2 = 15.70$). Species richness and Shannon's diversity index were 109 and 3.5 for the HDUZ, 141 and 3.7 for the LDUZ, and 96 and 3.8 for the natural forest, respectively.

Overall, there were more species per unit area in the HDUZ (0.25 ha^{-1}) compared to the LDUZ (0.18 ha^{-1}) and the neighboring natural forest (0.11 ha^{-1}). The native tree species composition was less than 50% in the HDUZ, slightly above 50% in the LDUZ, and about 95% in the natural forest (Figure 6). As a result, the HDUZ and LDUZ are more similar in composition, with high adjusted Jaccard's and Sørenson's indices, while the HDUZ and natural forest are the most dissimilar, with low indices and number of shared species (Table 4).

The tree density was higher in the HDUZ (366 trees ha^{-1}) than the LDUZ (351 trees ha^{-1}), while the sizes of trees in the LDUZ (DBH = 45.6 cm) were significantly larger ($p = 0.0144$) in girth and height compared to those of the HDUZ (DBH = 40.8 cm).

Table 4. Similarity (Jaccard and Sørenson indices) in species composition among urban zones in Kumasi. Values close to one indicate high similarity and close to zero indicate high dissimilarity. Bootstrap 95% confidence intervals are also shown.

Urban Zone	Shared Species	Jaccard Index			Sørenson $^\lambda$ Index		
			95% LL	95% UL		95% LL	95% UL
HDUZ–LDUZ	74	0.897	0.559	2.230	0.946	0.717	1.380
LDUZ–NF [1]	45	0.520	0.218	1.204	0.684	0.358	1.092
HDUZ–NF	27	0.312	0.135	0.605	0.476	0.239	0.754

[1] NF = Natural forest, HDUZ = high density urban zone, LDUZ = low density urban zone. $^\lambda$ LL = Lower limit, UL = Upper limit.

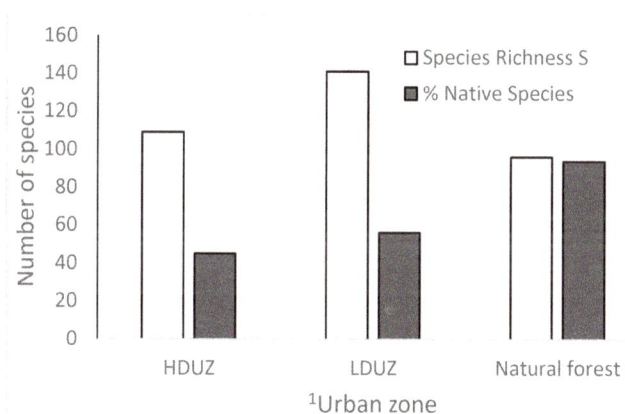

Figure 6. The species richness and proportion of native species in the urban zones in Kumasi.
[1] HDUZ—high density urban zone, LDUZ—low density urban zone and natural forest (Owabi sanctuary).

3.4. Carbon Storage among Diameter Classes

Although lower diameter classes had the highest tree densities and species diversity (richness), the amount of carbon sequestered was low compared to trees in the upper diameter classes (Figure 7). The carbon density of the trees in the city increases exponentially with the girth (size) of the trees. It ranges between 11 kg ha^{-1} for plants in the Arecaceae family, who's DBHs were not measured to over 500 kg ha^{-1} for the largest sized trees in the city. Overall, trees with DBH > 60 cm store over 54% of the total aboveground vegetation carbon (1.2 million t) of Kumasi (Figure 7). However, this diameter class constitutes only 17 and 9% of the total species richness and abundance of trees in the city, respectively. Areas with small-sized trees have a greater potential for carbon sequestration than areas with large-sized trees.

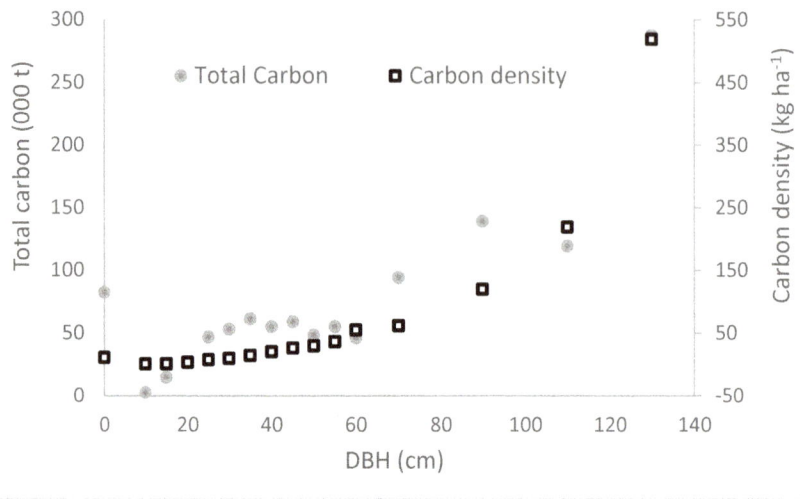

Figure 7. The total carbon stored and carbon stock densities for the different diameter classes in the urban forest of Kumasi.

4. Discussion

In this paper, spatially extensive data were used to identify variations in the structure, floristic, and functional composition of forest patches in Kumasi. These variations among green space types imply prominent differences in the population and ecosystem functions and call for different management and conservation strategies.

4.1. Species Diversity and Composition among Green Space Types

Inventory of the woody vegetation in Kumasi revealed a degree of species diversity among green spaces. Global urban forest inventory data is scarce because of the difficulty in operationalizing the concept [36]. The findings of this study represent perhaps the first comprehensive urban tree inventory in Ghana at a citywide scale. The species richness of 176 in Kumasi is numerically comparable or divergent to the species richness, diversity, and composition of other traditional land uses and national parks in Ghana. It is greater than the tree species richness of 73 in Kakum National Park [37] and 70 in the Boabeng-Fiama monkey sanctuary [38], and similar to the 171 species in the Tano Offin globally significant biodiversity area (GBSA) [39], as well as the 126 and 133 species found in the natural forest and fallow lands, respectively, within the high forest zone of Ghana [40]. Urban areas are comprised of modified abiotic and biotic environments which create room for species of native and exotic origins, as well as species of different guild compositions, to thrive [41,42]. This explains why the species richness and diversity of Kumasi is comparatively higher than other habitat types.

The overall species richness and Shannon index of 176 and 3.70, respectively, compare favorably well with tree diversity data of other cities in the region. In Abuja, Nigeria, the species richness and Shannon index of 69 and 3.56, respectively, were recently reported [43], while in Lome, Togo, 297 tree species in 141 genera and 48 families were recorded, with 69% of the species being alien [44]. Similarly, within the campus of the Valley View University, near Accra, 108 plant species in 51 families were recorded, of which 89% were native to the thickets of the Accra plains [45]. The high diversity and richness values in Kumasi conform with recent findings that cities in Africa support an extremely high diversity, even when compared to neighboring natural forest [43,44]. The relatively high presence of non-native species may be linked to the city's culture and history of development. Historically, when Kumasi was designated the "Garden city" of West Africa, several parks and green belts were created for agricultural production, entertainment (sports) for the British imperialists, and the enhancement of environmental quality [46]. In these green belts and parks, exotic vegetable and ornamental species were cultivated to beautify the residential areas and provide fresh food for the European settlers [47]. The inter-city differences in the diversity and composition of woody species may, however, be attributed to the morphology and age of the city, the natural ecological factors, and the building and population densities.

Among green spaces, there were significant differences in species diversity and composition in Kumasi. These differences reflect human management and stratified ecological conditions within the urban landscape. On farmlands and cemeteries, certain species may be selectively preserved because of the desired key services they confer on society. Natural forests and public parks are mostly protected areas, with the latter intended for recreational use. Consequently, these spaces are moderately disturbed and comprise native species such as *Triplichiton scleroxylon*, *Morinda lucida*, and a host of others. Vegetation near residential, institutional, and commercial buildings, as well as on streets, was dominated by a mix of species planted for their aesthetic, amenity, and alimentary values. It is worth noting that the top five most common species in the home gardens are all edible (Table 2). These findings agree with other studies which found that *Terminalia* spp., *Gmelina arborea*, *Tectona grandis*, *Delonix regia*, and *Elaeis guineensis* species were the most common species in urban areas in the tropical rainforest zone of West Africa [7]. The variations in species composition among the different green space types underscore the multidimensional attribute of the urban forest and its potential to provide multiple products and services to satisfy multiple societal and personal needs. These divergent forces underpinning the composition of the urban forest will have cascading effects on the urban forest structure.

4.2. The Forest Structure and Carbon Storage in Green Spaces of Kumasi

The forest structure is described by its tree density, basal area, and height and DBH class distributions. The mean tree density and basal area of the urban forest of Kumasi somewhat deviated from what is commonly reported for natural and plantation forests around the country. The mean basal area and density of trees in the rain forest of Ghana are 23.8 m^2 ha^{-1} and 487 trees ha^{-1}, respectively [48]. This deviation may be attributed to the assorted conservation or protection of large girthed trees in natural forest relics, cemeteries, public parks, institutional compounds, and farmlands in Kumasi. However, the tree density of Kumasi is similar to the 141 trees $acre^{-1}$ (348 trees ha^{-1}) in Gainesville, Florida, USA [6]. Basal area and tree density differed among green space types (Table 3). These differences reflect the degree of stress, anthropogenic disturbance, and history of urban development. Whereas natural forest and public parks may be maintained as default establishment of tree species with limited stress and human interference, in green spaces such as home gardens, institutional compounds, streets, and plantations, desired species for specified purposes are usually planted to meet individual and community amenity values. The local government in Kumasi recently launched a project to plant over one million trees along streams and major streets in the city. Trees along streets and near residential areas are more susceptible to stress, hence are more dynamic in population and structure [10].

In many studies, the reverse J-shaped model perfectly describes the structure of the tropical forest in Africa [49,50]. Ecologically, the reverse J-shaped model is ascribed to equal mortality among the various diameter classes. In Kumasi, the diameter class distribution of the trees followed a modified reverse J-shaped curve. Furthermore, while in some green spaces, the actual reverse J-shaped curve applied, in some others, it did not. These trends in Kumasi can be attributed to the complex combination of forest and non-forest tree species in the different green space types, as well as the presence of natural and plantation forest types. The findings in Kumasi are, however, consistent with others which posited that the reverse J-shaped curve is not the only model that describes the forest structure [51].

At a citywide scale, it was found that species richness and diversity decreased with increasing diameter class. This is consistent with the findings in a Madagascar rainforest, which showed that diversity and richness decreased with the increasing girth of trees [52]. The forest structure of Kumasi, to a large extent, is similar to that of neighboring forest, i.e., showing similarities in diameter, height, and basal area distribution. Similarly, the carbon storage in trees increased with increasing diameter class in Kumasi, although large-sized trees were comparatively fewer than small-sized trees. Carbon storage is a function of size (girth), wood density, and height [29,53]. Small-sized trees, therefore, store less carbon compared to large-sized trees. Although evidence from the tropical forest indicates no relationship between diversity and forest carbon storage [54], data in Kumasi seem to show that tree species diversity within diameter classes may be inversely related to carbon storage. The present study did not categorically estimate other climate mitigation services, such as evaporative cooling and shading resulting from urban tree cover, as chronicled in other studies [55–57]. Nonetheless, it can be inferred that, given the wide diversity of these broad-leaf tree species in Kumasi, the urban forest would be contributing massively to mitigating heat island effects. The findings highlight the importance of the urban landscape in biodiversity conservation and enhancing ecosystem function at the city, national, and regional scales.

4.3. Implications for Conservation

Some species listed by the IUCN as vulnerable and endangered are recorded in the forest of Kumasi. These include *Adansonia digitata*, *Afzelia africana*, *Albizia ferruginea*, *Antiaris toxicaria*, *Chlorophora excelsa*, *Entandrophragma* spp., *Hallea* spp., *Hymenostegia aubrevillei*, *Khaya senegalensis*, *Prosopis africana*, *Rauwolfia vomitoria*, *Samanea dinklagei*, and *Terminalia ivorensis* [58]. Eighteen important Ghanaian timber species were also recorded in the city. These include *Chlorophora excelsa*, *Antiaris toxicaria*, *Entandrophragma* spp., *Albizia* spp., *Pycnantus angolensis*, *Ceiba pentandra*, and a host of others. These were mostly in the upper diameter (>60 cm) classes. Only *Albizia zygia*, *Alstonia boonei*, and *Mammea africana* were found in the lower diameter (<20 cm) classes. This is in accordance with studies in cities which concluded that urban green spaces are repositories of biological diversity, of which at least 8% native bird and 25% native plant species are present [59]. Forests in/near urban areas are targeted for conservation because they harbor biodiversity, including IUCN vulnerable and endangered species [60]. Indeed, significant numbers of birds, mammals, reptiles, amphibians and invertebrates have been reported near some urban centers in Ghana [61]. These show that urban green spaces provide habitats for trees and several fauna species.

In Kumasi, the presence of large-sized native tree species in the natural forest, cemeteries, public parks, and farmlands mimics old-growth forests. Old-growth forests are frequently targeted for conservation since they harbor a large proportion of vulnerable species (disturbance sensitive) and species of restricted distribution [49]. In addition, the high abundance of small-sized trees in Kumasi suggest its urban forest, if maintained and managed adequately, can be sustainable. The wide variation in the structural attributes of the urban forest of Kumasi, as well as patch sizes, is fundamental to supporting high biological diversity and hence warrants being conserved. However, considering that urban biodiversity is determined to a large extent by human preferences, high tree diversity may not imply high overall faunal diversity [62]. To enhance the biological diversity of Kumasi and other urban areas further, the following suggestions may be pursued:

1. Deliberate efforts to expand the urban forest in Kumasi and indeed across Ghana and Africa ought to be pushed forward. This requires planting and conserving a wide variety of native species as well as well-adapted economically-beneficial exotic species, especially in the HDUZ (core) areas of the city.

2. Although invasive species are currently not studied within Kumasi, it is worth noting that some exotic species could potentially be invasive. Therefore, practitioners should judiciously select species when implementing urban afforestation programs.
3. Grasslands and farmlands which are commonly found along wetlands in Kumasi could be planted with trees of assorted species mixes and excluded or protected from build-up development. In this regard, a form of urban agroforestry could be practiced, while at the same time helping urbanites adhere to the 100 m no development zones near water bodies in urban areas in Ghana.
4. Biodiversity in public parks and wetlands within Kumasi could be further guided and protected by formidable policies. This will perhaps require bye-laws enacted by the metropolitan authorities in conjunction with the communities.
5. Considering that home gardens support a high wealth of diversity and are at risk of being cleared for further housing development as and when the economic situation of the land owner permits, it is suggested that clear policy directions be enacted, demanding every parcel of land designated for housing to maintain a certain minimum amount of green cover and tree cover.
6. Both in situ and ex situ conservation measures ought to be adopted to conserve indigenous species in these landscapes, especially rare species and/or shade-tolerant species.
7. Stakeholders in urban forest (ry) need to explore better ways of making urban biodiversity profitable in order to generate income and enhance environmental sustainability. Examples are adhering to best management practices within no-development zones along waterbodies and using these riverine corridors for tourism to generate income.

5. Conclusions

Urban forestry is emerging as a means to enhance urban environmental sustainability and resilience, especially in developing countries. This study discusses the structure and diversity in relation to carbon storage of the urban forest of Kumasi. The species richness and diversity of the city is similar and sometimes higher than that of other landscapes and national parks within the same ecoregion in the country. Structurally, a modified reverse J-shaped model described the distribution of trees in the city and varied among green space types. The modification was due to the inclusion of palms and non-traditional timber species that are a rudimentary part of the urban forest. It is concluded that the urban forest structure is unique and different from that of other forest types.

While species richness and diversity decreases with increasing diameter class, carbon storage increases with increasing diameter class. The selective preservation of certain species and natural processes in the urban spaces accounts for these trends. Carbon storage is a function of tree size and not necessarily tree density. Hence, a few large trees stored more carbon than a bunch of small trees. These findings provide baseline information about the forest structure and species composition and should be the basis for urban planning decision making regarding green urbanism in Kumasi and other cities in Ghana and Africa.

Integrated and coordinated efforts by the local government, urban planners, traditional leaders, ecologists, and environmentalists, as well as the general public, ought to be harnessed to facilitate adequate conservation and utilization of urban forest towards coping with environmental hazards, minimizing poverty, and increasing food security. Further research could be tailored towards the structure, diversity, and function of urban forest in relation to poverty reduction, food security, and public health. Further research should also emphasize tree compositional and structural relationships with microclimatic regulation and shade benefits within urban landscapes in cities in Africa.

Author Contributions: B.F.N. designed the study, collected the data, and wrote the paper; D.C.-C. and M.D. provided guidance and supervision. All authors reviewed the paper.

Funding: The study was funded by Foundation Fiat Panis, the German Federal Ministry of Economic Corporation via the German Academic Exchange Service (DAAD), and the German Federal Ministry of Education and Research (BMBF) through the BiomassWeb project.

Acknowledgments: I am greatly indebted to the funders of this study. I am equally grateful to numerous persons who assisted in many diverse ways. Many thanks to Jonathan Dabo of CSIR-FORIG, Degan Amissah, formerly of the KNUST Department of Biology and Asare of KNUST Faculty of Pharmacy, for assisting with identification of the flora. To Bertrand Yosangfo, Callistus Nero, and Isaac Boateng, your assistance with vegetation data collection is greatly appreciated. We are grateful to two anonymous reviewers for their useful comments and suggestions.

Conflicts of Interest: Authors declare no conflicts of interest.

References

1. Ghana Statistical Service. *2010 Population and Housing Census Final Results Ghana Statistical Service*; Ghana Statistical Service: Accra, Ghana, 2012.
2. United Nations. *World Urbanization Prospects*; United Nations: New York, NY, USA, 2014.
3. Nero, B.F. Urban green space dynamics and socio-environmental inequity: Multi-resolution and spatiotemporal data analysis of Kumasi, Ghana. *Int. J. Remote Sens.* **2017**, *38*, 6993–7020. [CrossRef]
4. Nero, B.F. *Urban Green Spaces Enhance Carbon Sequestration and Conserve Biodiversity in Cities of the Global South: Case of Kumasi Ghana*; University of Bonn: Bonn, Germany, 2017.
5. Konijnendijk, C.C.; Ricard, R.M.; Kenney, A.; Randrup, T.B. Defining urban forestry—A comparative perspective of North America and Europe. *Urban For. Urban Green.* **2006**, *4*, 93–103. [CrossRef]
6. Escobedo, F.; Seitz, J.A.; Zipperer, W. *Gainesville's Urban Forest Structure and Composition*; University of Florida IFAS Extension: Gainesville, FL, USA, 2009; pp. 1–3.
7. Fuwape, J.A.; Onyekwelu, J.C. Urban Forest Development in West Africa: Benefits and Challenges. *J. Biodivers. Ecol. Sci.* **2011**, *1*, 77–94.
8. Sanders, R.A. Some determinants of urban forest structure. *Urban Ecol.* **1984**, *8*, 13–27. [CrossRef]
9. Mcbride, J.R.; Jacobs, D.F. Urban forest structure: A key to urban forest planning. *Calif. Agric.* **1979**, *33*, 24.
10. Sæbø, A.; Benedikz, T.; Randrup, T.B. Selection of trees for urban forestry in the Nordic countries. *Urban For. Urban Green.* **2003**, *2*, 101–114. [CrossRef]
11. Buckelew Cumming, A.; Nowak, D.J.; Twardus, D.B.; Hoehn, R.E.; Mielke, M.; Rideout, R. *The Urban Forest of Wisconsin 2002—Pilot Monitoring Project 2002*; United States Forest Service: Washington, DC, USA, 2007.
12. Escobedo, F.J.; Nowak, D.J.; Wagner, J.E.; Luz de la Maza, C.; Rodriguez, M. The socioeconomics and management of Santiago de Chile urban forest. *Urban For. Urban Green.* **2006**, *4*, 105–114. [CrossRef]
13. Nowak, D.J.; Crane, D.E.; Stevens, J.C.; Hoehn, R.E.; Walton, J.T.; Bond, J. A Ground-Based Method of Assessing Urban Forest Structure and Ecosystem Services. *Aboric. Urban. For.* **2008**, *34*, 347–358.
14. Nowak, D.J.; Crane, D.E. Carbon storage and sequestration by urban trees in the USA. *Environ. Pollut.* **2002**, *116*, 381–389. [CrossRef]
15. Nielsen, A.B.; Östberg, J.; Delshammar, T. Review of urban tree inventory methods used to collect data at single-tree level. *Arboric. Urban. For.* **2014**, *40*, 96–111.
16. Alonzo, M.; Mcfadden, J.P.; Nowak, D.J.; Roberts, D.A. Mapping urban forest structure and function using hyperspectral imagery and lidar data. *Urban For. Urban Green.* **2016**, *17*, 135–147. [CrossRef]
17. Davies, Z.G.; Edmondson, J.L.; Heinemeyer, A.; Leake, J.R.; Gaston, K.J. Mapping an urban ecosystem service: Quantifying above-ground carbon storage at a city-wide scale. *J. Appl. Ecol.* **2011**, *48*, 1125–1134. [CrossRef]
18. Kim, G. Assessing Urban Forest Structure, Ecosystem Services, and Economic Benefits on Vacant Land. *Sustainability* **2016**, *8*, 679. [CrossRef]
19. Schreyer, J.; Tigges, J.; Lakes, T.; Churkina, G. Using airborne LiDAR and QuickBird data for modelling urban tree carbon storage and its distribution—A case study of Berlin. *Remote Sens.* **2014**, *6*, 10636–10655. [CrossRef]
20. Food and Agriculture Organization of the United Nations (FAO). *Urban and Peri-Urban Forestry in Africa: The Outlook for Woodfuel*; FAO: Rome, Italy, 2012.
21. Department for International Development (DFID). *Forest Inventory Project, Ghana*; DFID: London, UK, 2004.
22. Treue, T. National Forest Inventory Continued (1989–1997). In *Politics and Economics of Tropical High Forest Management*; Kluwer Academic Publishers: South Holland, The Netherlands, 2001; pp. 37–38.
23. Owusu, K. *Changing Rainfall Climatology of West Africa: Implications for Rainfed Agriculture in Ghana and Water Sharing in the Volta Basin*; University of Florida: Gainesville, FL, USA, 2009.

24. Manu, A.; Twumasi, Y.A.; Coleman, T.L. Is It the Result of Global Warming or Urbanization? The Rise in Air Temperature in Two Cities in Ghana. In Proceedings of the 5th FIG Regional Conference, Accra, Ghana, 8–11 March 2006.
25. Kumasi Metropolitan Assembly (KMA). *The Composite Budget of the Kumasi Metropolitan Assembly (KMA) for the 2013 Fiscal Year*; KMA: Kumasi, Ghana, 2013.
26. Adarkwa, K.K. The role of Kumasi in national development—Kumasi as a central place. In *Future of the Tree: Towards Growth and Development of Kumasi*; Adarkwa, K.K., Ed.; University Printing Press: Kumasi, Ghana, 2011; pp. 14–34.
27. Hawthorne, W.; Gyakari, N. *Photoguide for the Forest Trees of Ghana*; Trees Spotters's Field Guide for Identifying the Largest Trees; Oxford Forestry Institute: Oxford, UK, 2006.
28. Oteng-Amoako, A.A. *100 Tropical African Timber Trees from Ghana: Tree Description and Wood Identification with Notes on Distribution, Ecology, Silviculture, Ethnobotany, and Wood Uses*; Forest Research Institute of Ghana: Kumasi, Ghana, 2002.
29. Chave, J.; Réjou-Méchain, M.; Búrquez, A.; Chidumayo, E.; Colgan, M.S.; Delitti, W.B.C.; Duque, A.; Eid, T.; Fearnside, P.M.; Goodman, R.C.; et al. Improved allometric models to estimate the aboveground biomass of tropical trees. *Glob. Chang. Biol.* **2014**, *20*, 3177–3190. [CrossRef] [PubMed]
30. Food and Agriculutre Organization of the United Nations (FAO). *Estimating Biomass and Biomass Change of Tropical Forests*; FAO: Rome, Italy, 1997.
31. Orwa, S.; Mutua, A.; Kindt, R.; Jamnadass, R.H.; Simons, A. *Agroforestree Database: A Tree Reference and Selection Guide, Version 4.0*; World Agroforestry Center: Nairobi, Kenya, 2009.
32. Martin, A.R.; Thomas, S.C. A reassessment of carbon content in tropical trees. *PLoS ONE* **2011**, *6*. [CrossRef] [PubMed]
33. Chao, A.; Chazdon, R.L.; Colwell, R.K.; Shen, T.J. Abundance-based similarity indices and their estimation when there are unseen species in samples. *Biometrics* **2006**, *62*, 361–371. [CrossRef] [PubMed]
34. Chao, A.; Chazdon, R.L.; Shen, T.J. A new statistical approach for assessing similarity of species composition with incidence and abundance data. *Ecol. Lett.* **2005**, *8*, 148–159. [CrossRef]
35. Koleff, P.; Gaston, K.J.; Lennon, J.J. Measuring Beta Diversity for Presence-Absence Data. *J. Anim. Ecol.* **2003**, *72*, 367–382. [CrossRef]
36. Konijnendijk, C.C.; Gauthier, M. *Urban Forestry for Multifunctional Urban Land Use*; RUAF Foundation: Leusden, The Netherlands, 2014.
37. Pappoe, A.N.M.; Armah, F.A.; Quaye, E.C.; Kwakye, P.K.; Buxton, G.N.T. Composition and stand structure of a tropical moist semi-deciduous forest in Ghana. *Plant Sci.* **2010**, *1*, 95–106.
38. Kankam, B.O.; Saj, L.; Sicotte, P. Short-term variation in forest dynamics: Increase in tree diversity in Boabeng-Fiama, monkey sanctuay, Ghana. *Ghana J. For.* **2013**, *29*, 19–33.
39. Enninful, R. *Assessment of Floral Compostion, Structure and Natural Regeneration of the Tano Offin Globally Significant Biodiversity Area*; Kwame Nkrumah University of Science and Technology: Kumasi, Ghana, 2013.
40. Anglaaere, L.C.N.; Cobbina, J.; Sinclair, F.L.; McDonald, M.A. The effect of land use systems on tree diversity: Farmer preference and species composition of cocoa-based agroecosystems in Ghana. *Agrofor. Syst.* **2011**, *81*, 249–265. [CrossRef]
41. McKinney, M.L. Urbanization as a major cause of biotic homogenization. *Biol. Conserv.* **2006**, *127*, 247–260. [CrossRef]
42. McKinney, M.L. Effect of urbanization on species richness: A review of plants and animals. *Urban Ecosyst.* **2008**, *11*, 161–176. [CrossRef]
43. Agbelade, A.D.; Onyekwelu, J.C.; Oyun, M.B. Tree Species Richness, Diversity, and Vegetation Index for Federal Capital Territory, Abuja, Nigeria. *Int. J. For. Res.* **2017**, *2017*. [CrossRef]
44. Raoufou, R.; Kouami, K.; Koffi, A. Woody plant species used in urban forestry in West Africa: Case study in Lomé, capital town of Togo. *J. Hortic. For.* **2011**, *3*, 21–31.
45. Simmering, D.; Addai, S.; Geller, G.; Otte, A. A university campus in peri-urban Accra (Ghana) as a haven for dry-forest species. *Flora Veg. Sudano-Sambesica* **2013**, *16*, 10–21.
46. Quagraine, V.K. Urban landscape depletion in the Kumasi Metropolis. In *Future of the Tree: Towards Growth and Development of Kumasi*; Adarkwa, K.K., Ed.; University Printing Press: Kumasi, Ghana, 2011; pp. 212–233.
47. Asomani-Boateng, R. Urban Cultivation In Accra: An Examination of The Nature, Practices, Problems, Potentials And Urban Planning Implications. *Habitat Int.* **2002**, *26*, 591–607. [CrossRef]

48. Hall, J.B.; Swaine, M.D. Classification and Ecology of Closed-Canopy. *J. Ecol.* **1976**, *64*, 913–951. [CrossRef]
49. Fayolle, A.; Swaine, M.D.; Bastin, J.-F.; Bourland, N.; Comiskey, J.; Dauby, G.; Doucet, J.-L.; Gillet, J.-F.; Gourlet-Fleury, H.S.; Kirunda, O.J.B.; et al. Patterns of tree species composition across tropical African forests. *J. Biogeogr.* **2014**, *41*, 2320–2331. [CrossRef]
50. Kacholi, D.S. Analysis of Structure and Diversity of the Kilengwe Forest in the Morogoro Region, Tanzania. *Int. J. Biodivers.* **2014**, *2014*. [CrossRef]
51. Westphal, C.; Tremer, N.; Von Oheimb, G.; Hansen, J.; Von Gadow, K.; Ha, W. Is the reverse J-shaped diameter distribution universally applicable in European virgin beech forests? *For. Ecol. Manag.* **2006**, *223*, 75–83. [CrossRef]
52. Armstrong, A.H.; Shugart, H.H.; Temilola, E. Characterization of community composition and forest structure in a Madagascar lowland rainforest. *Trop. Conserv. Sci.* **2011**, *4*, 428–444. [CrossRef]
53. Chave, J.; Andalo, C.; Brown, S.; Cairns, M.; Chambers, J.Q.; Eamus, D.; Fölster, H.; Fromard, F.; Higuchi, N.; Kira, T.; et al. Tree allometry and improved estimation of carbon stocks and balance in tropical forests. *Oecologia* **2005**, *145*, 87–99. [CrossRef] [PubMed]
54. Sullivan, M.J.P.; Talbot, J.; Lewis, S.L.; Phillips, O.L.; Qie, L.; Begne, S.K.; Chave, J.; Cuni-Sanchez, A.; Hubau, W.; Lopez, G.; et al. Diversity and carbon storage across the tropical forest biome. *Sci. Rep.* **2017**, *7*. [CrossRef] [PubMed]
55. Rahman, M.A.; Moser, A.; Gold, A.; Rötzer, T.; Pauleit, S. Vertical air temperature gradients under the shade of two contrasting urban tree species during different types of summer days. *Sci. Total Environ.* **2018**, *633*, 100–111. [CrossRef] [PubMed]
56. Rahman, M.A.; Armson, D.; Ennos, A.R. A comparison of the growth and cooling effectiveness of five commonly planted urban tree species. *Urban Ecosyst.* **2015**, *18*, 371–389. [CrossRef]
57. Moser, A.; Rahman, M.A.; Pretzsch, H.; Pauleit, S.; Rötzer, T. Inter- and intraannual growth patterns of urban small-leaved lime (*Tilia cordata mill.*) at two public squares with contrasting microclimatic conditions. *Int. J. Biometeorol.* **2017**, *61*, 1095–1107. [CrossRef] [PubMed]
58. International Union for Conservation of Nature (IUCN). *IUCN Red Lists of Threatened Species*; IUCN: Gland, Switzerland, 2013.
59. Aronson, M.F.J.; La Sorte, F.A.; Nilon, C.H.; Katti, M.; Goddard, M.A.; Lepczyk, C.A.; Warren, P.S.N.; Williams, S.G.; Cilliers, S.; Clarkson, B.; et al. A global analysis of the impacts of urbanization on bird and plant diversity reveals key anthropogenic drivers. *Proc. R. Soc. B Biol. Sci.* **2014**, *281*. [CrossRef] [PubMed]
60. Bulafu, C.; Baranga, D.; Mucunguzi, P.; Telford, R.J.; Vandvik, V. Massive structural and compositional changes over two decades in forest fragments near Kampala, Uganda. *Ecol. Evol.* **2013**, *31*, 3804–3823. [CrossRef] [PubMed]
61. Deikumah, J.P.; Kudom, A.A. Biodiversity Status of Urban Remnant Forests in Cape Coast, Ghana. *J. Sci. Technol.* **2010**, *30*, 1–8. [CrossRef]
62. Faeth, S.H.; Bang, C.; Saari, S. Urban biodiversity: Patterns and mechanisms. *Ann. N. Y. Acad. Sci.* **2011**, *1223*, 69–81. [CrossRef] [PubMed]

© 2018 by the authors. Licensee MDPI, Basel, Switzerland. This article is an open access article distributed under the terms and conditions of the Creative Commons Attribution (CC BY) license (http://creativecommons.org/licenses/by/4.0/).

Article

Preferences of Tourists for the Service Quality of Taichung Calligraphy Greenway in Taiwan

Wan-Yu Liu [1,2,*] and Ching Chuang [1,*]

1. Department of Forestry, National Chung Hsing University, Taichung 402, Taiwan
2. Innovation and Development Center of Sustainable Agriculture, National Chung Hsing University, Taichung 402, Taiwan
* Correspondence: wyliu@nchu.edu.tw (W.-Y.L.); ching41041@gmail.com (C.C.); Tel.: +88-642-285-0158 (W.-Y.L.)

Received: 12 July 2018; Accepted: 25 July 2018; Published: 30 July 2018

Abstract: This study explores preferences for a set of attributes that characterize the recreational value of Calligraphy Greenway, the most notable greenbelt in Taichung City, Taiwan. As an urban green space, the Calligraphy Greenway has its own recreational attributes and visitors' preferences. This study uses the choice experiment method to determine visitors' preference levels for five major attributes to improve the recreational quality. On average, each visitor visited there 9.15 times in the past year and spent 2.37 h per visit. Of the five recreational attributes, satisfaction with recreational activity opportunities had the highest score and satisfaction with cultural landscape resources had the lowest score. The importance is ranked in the order of recreational service quality, total recreational cost, environmental landscape resources, cultural landscape resources and recreational activity opportunities. Considering difference of groups, female visitors were more concerned with cost and activities but male visitors were more concerned with service quality and natural/cultural landscape resources. Local visitors were more concerned with cost and activities but non-local visitors were more concerned with environmental/cultural landscape resources. Both were concerned with service quality. Based on the results, this study makes the following recommendations: cultural landscape resources and quality of recreational services and facilities should be improved and more complete interpretative educational guidance should be provided to increase visitors' willingness to visit. Additionally, it is suggested to set up various districts to cater for preferences of different visitor groups.

Keywords: Greenway; urban forest; preferences; choice experiment

1. Introduction

The advancement of human civilization means the development of urban functions and the need for a better balance between work and recreation. Therefore, recreation now occupies an increasingly larger proportion of daily life. This has led to a greater demand for recreation, which drives urban planning and puts a greater value on urban forests and green space. Urban forests and green spaces are critical for social, environmental and economic benefits, along with green infrastructure. They are increasingly considered to be an important part of a city's image [1–3]. Gobster and Westphal [2] investigated how people perceived and used greenways for recreation and experiences of relaxation and their results accounted for the importance of greenways by using six human dimensions in Chicago, USA. Weber et al. [3] stated that urban greenways would have impacts on the residents in various perspectives, including property values, social and recreation spaces and so on. They surveyed 381 adjacent residents in BeltLine, Atlanta, USA to investigate the factors and preferences for planning urban greenways. The green space in cities provides citizens with a space that is suited for moderate exercise to relieve the stresses of daily life and to enjoy the beauty of a forest. Previous studies have shown that leisure activities are an essential part of modern life [4,5].

In Taiwan, a city's degree of greenery is a factor in evaluating a city's level of development. Taichung's post amalgamation green coverage exceeds 57.7%, with a total of 9.62 m^2 of parks, green areas, squares and playgrounds per person, which ranks first among Taiwan's five municipalities (Taipei, Xinbei, Taichung, Tainan and Kaohsiung). Taichung has made its greenway plan since 1931 [6]. As the city grew, the original design changed with the times and the functions of greenways shifted. The greenways in Taiwanese were originally geared towards facilitating liaison and gradually became to include more recreational and artistic functions. In 2008, Taichung featured a total of 13 greenways, as shown in Figure 1. In recent years, Taichung's city government has implemented a series of greenway plans and the "Calligraphy Greenway Plan" is the most important one [7,8].

Figure 1. *Cont.*

Figure 1. Taichung Greenways.

Taiwan has a subtropical monsoon climate [9]. With such a geographical location and a colonial history, the population composition in Taiwan is multicultural. Taichung is located in the middle of the Taiwan island and is the transportation hub between Southern and Northern Taiwan. Hence, this study provides a valuable reference for designing and maintaining urban green space and forests in the similar climate zones and multicultural cities. Because the preferences of urban residents are associated with the whole residents' diversity of cultural backgrounds and the environments of subtropical climate, they not only provide policy and management considerations to local governments but also provide a reference of parameter values for management design of urban forests.

When constructing the Taichung Calligraphy Greenway, the government adopted eco-friendly materials and took the whole area into consideration [10]. For instance, the major shopping mall in the Calligraphy Greenway area is covered by a vertical garden composed of hundreds of thousands of plants, which is known as the largest green wall in Asia. The building received the 2010 FIABCI Prix d'Excellence Awards [11]. The Taichung Calligraphy Greenway also includes a bicycle lane with YouBike (the largest public bicycle sharing system in Taiwan) stations for renting shared bicycles. When designing the Calligraphy Greenway, the goal was to provide a low-carbon recreational environment and to enhance the city's image [8]. In order to reduce the carbon emissions, the Taichung government reshaped the living space and negotiated with the neighboring private industries.

Taichung Calligraphy Greenway is a significant attraction in Taichung. It not only provides a recreation area to adjacent residents but also appeals visitors from other places. Therefore, the Taichung government eagers to gather more information for improving the quality of the Taichung Calligraphy Greenway in order to provide the residents a better recreation area and to promote the eco-friendly urban recreation via the greenway. Thus, investigating the visitors' latent preferences could assist the government to establish better management and planning. In order to investigate the visitors' preferences, previous studies showed that combinations of recreational attributes and choices impacted tourists' decisions [12] and affected various attributes of recreational zone installations. Among several methods, the method based on economic evaluation treated the abstract value of each attribute as a monetary value and a change in the levels of these attributes showed the recreational preferences [4]. Various factors (such as the region, culture, social demographic structure, economic background and recreational habits) have been studied [5].

This study uses choice experiment (CE) to determine Taichung citizens' preferences in terms of the Calligraphy Greenway's attributes. CE have been widely used in economic evaluations of forest recreation [13,14]. When testing CE, the respondent chooses an alternative group, which actually represents a product or a service change value. The respondent's choice reflects preferences [15]. The CE was proved effective especially for the recreational evaluation of urban forests or green zones [4]. The goal of this study is to analyze (1) the effect of visitors' preference levels to Taichung Calligraphy Greenway on different recreational attributes and (2) the effect of preference levels of visitors with different socioeconomic backgrounds on different recreational attributes. The results could provide a valuable reference for planning other city greenways.

2. Material and Methods

2.1. Theoretical Model

CE is originated from Lancaster's consumer theory, which states that the utility of a commodity is the total sum of utilities of all its attributes [16]. In the theory, each attribute of a commodity affects consumer choice. Therefore, in a CE test, a combination of multiple attributes of a commodity constitute a bundle of attributes.

According to the Random Utility Model theory (RUM) [17], the utility U_{ij} of a product is a random variable determined by a deterministic component V_{ij} (representing the product's attributes that can be directly observed) and a random component ε_{ij}. In general, because there are too many combinations of all levels of attributes, only a part of these combinations is provided to respondents in the survey. Then, the utility U_{ij} that respondent i obtains from selecting alternative j is represented as follows:

$$U_{ij} = V_{ij} + \varepsilon_{ij} = \beta_i x_{ij} + \varepsilon_{ij} \tag{1}$$

where i indicates the index of a respondent; j indicates the index of the alternative that the respondent chooses from the given choice set; V_{ij} is the component that can be directly observed when respondent i chooses alternative j, also being expressed as $\beta_i x_{ij}$, in which β_i is a coefficient vector and x_{ij} is a vector of attribute values of alternative j chosen by respondent i [17,18]; and ε_{ij} indicates unobservable attributes.

This model is based on the assumption that all respondents share common likes. However, this study aims to set group tests and different alternatives to infer preferences of the respondents. Therefore, this study uses the latent-class model (LCM) to verify the accuracy of the original hypothesis. The LCM supposes that the respondents are divided into C classes. Conditional Class-Membership Probabilities are not restricted to estimating coefficients but can also be used to define marginal values of attributes. Change of different levels of attributes determines "Willingness to Pay" (WTP). According to [18], the WTP of respondent i in class c is expressed as follows:

$$WTP_{i|c} = -\frac{1}{\beta_{payment,c}} \left[\ln\left(\sum_{k=1}^{I} \exp(\beta'_c x_k^0) \right) - \ln\left(\sum_{k=1}^{I} \exp(\beta'_c x_k^1) \right) \right] \tag{2}$$

where I denotes the set of indices of all the given alternatives; x_k^0 and $x_k^1 x_k^0$ represent vectors of attribute values of alternative k chosen by respondent i before and after change, respectively; x_k^1 β'_c is the coefficient vector for class c; and $\beta_{payment,c}$ is the payment coefficient for class c. This equation shows that the attribute bundles are swapped to calculate the possible marginal utility following the exchange, demonstrating the importance of attributes to the respondent.

To determine the utility of attributes, this study uses the stated preference method (a.k.a, laboratory simulation method or scenario-based method), in which real conditions are simulated by experimental designs and then the respondent makes a policy via a cognitive process [6]. The method has the following characteristics: (1) each hypothetical alternative is demonstrated to respondents through a description; (2) the description of each alternative is interpreted using a portion of the attributes of known or existing products or services; (3) each attribute has different levels and each alternative consists of a combination of attribute levels; (4) alternatives consisting of levels of attributes are designed by experimental designs; and (5) respondents use different methods to express their preferences for alternatives.

The information that influences respondents' choices can be ascertained using the stated preference method and hence alternatives are considered according to respondents' different backgrounds in the descriptive preference mode. Given a choice set of 2–5 alternatives, each respondent selects the preferred alternative from the set. Attributes are chosen to be independent of each other, to avoid structural mode failure because of linear relationships between attributes [4,15].

The alternatives in the stated preference method are described using the following three ways: textual description, paragraph description and graphic expression. Textual description uses texts to state attributes of each alternative and their levels and is simple, concise and highly efficient. Paragraph description describes hypothetical stimuli in a paragraph, so the entire circumstances are described to respondents. However, it can only be used to describe a limited number of stimuli. Graphic expression is the truest among the three methods but is costlier in terms of time and money. In order to maximize the questionnaire benefits under the limited resources, this study adopts textual description. The questionnaire given to respondents presented corresponding textual descriptions, to give respondents a basic understanding of the questionnaire's response method and alternative selection when completing the questionnaire.

Most of the studies on preference measurement used different measurement methods for different attributes, respondents, or experimental goals. Preference measurement methods are generally categorized as first preference method, ranking method, or rating methods. For the first preference method, each respondent chooses an alternative to replace the current alternative from all the simulated alternatives and the chosen alternative is regarded as the respondent's first preference. For the ranking method, each respondent ranks all alternatives according to the respondent's preference. For the rating method, each respondent sets a rate for each alternative, in which a higher rate indicates a greater preference for the alternative. Most European and US studies that applied conjoint analysis were based upon ranking or rating methods [19]. Previous studies to determine the validity of the two methods did not reach a uniform conclusion. Therefore, this study uses the most commonly used method: the rating method.

2.2. Determining Attributes that Impact Preferences and Their Levels

There are many ways to determine recreational attributes that influence preferences, such as the researcher's professional judgment [20], investigations by a group of professionals [21], or in-depth interviews with tourists [22]. The first part of is study refers to the analysis of the socio demographic data in order to categorize the demographic background of the visitors to the Calligraphy Greenway.

This study refers previous studies on Taichung Calligraphy Greenway [23] to organize the attributes that influence preferences, as shown in Table 1, in which the levels of each attribute are obtained by referring to previous studies and our discussions. Among the five attributes that influence recreational preference in Table 1, "total recreational cost" is regarded as an interval scale variable, which is a kind of statistic scale in which the value between the scale points is measurable (i.e., it has a linear decreasing relationship with the utility function so the higher the cost is, the lower the utility is). All the other four attributes are regarded as discrete variables. The "recreational service quality" (including quality of facilities), "environmental landscape resources" and "cultural landscape resources" are considered as ordinal scale variables. The higher the "recreational service quality" is, the higher the utility is. The more abundant the "environmental landscape resources" and "cultural landscape resources" are, the greater the utility is.

Table 1. The attributes that influence the preferences on selecting recreational areas and their levels.

Attribute	Description	Level
Total recreational cost	Transportation, accommodation, entertainment, souvenirs and opportunity costs that are associated with traveling to a recreational area [5,13,20].	1. 16.35 USD or lower 2. 16.35 USD or higher, 32.69 USD or lower 3. 32.69 USD and above
Recreational service quality	Recreational service facilities include paths, rest tables and chairs, lighting, pavilions, bike racks and garbage cans [13,23].	1. High quality 2. Medium quality 3. Low quality
Recreational activity opportunities	Opportunities of providing the ways in which tourists participate in activities, including observational activities (including jazz music, street art performances and activities that audiences passively observe) and participation activities (including live interactive games and green experiences) [8,23].	1. Only observational activities are provided 2. Only participation activities are provided 3. Both observational and participation activities are provided
Environmental landscape resources	Flora, including tall trees, shrubs and sod, in which tree species are primarily large-leaf mahogany, blackboard trees, Bauhinia Japonica, Royal Poinciana, Madagascar Almond and the floss-silk tree [23,24].	1. Abundant 2. Few
Cultural landscape resources	Landscapes or facilities that are related to culture, including public art and information signs [23].	1. Abundant 2. Few

Resource from: [20]; Conducted by this study.

2.3. Determining Alternatives

After the five attributes and their levels are determined, there are a total of $3^3 \times 2^2 = 108$ different alternatives if a complete factorial design is applied. If the questionnaire required respondents to conduct preference evaluations for these 108 alternatives, the questionnaire's evaluation tasks would be overly complicated, which could influence accuracy of the results. The number of alternatives can be reduced by using an orthogonal fractional factorial design (i.e., without considering the interaction between factors). The alternative sets were then determined by the statistic software and the experts [20]. By conducting SPSS (version 20.0, IBM, Armonk, NY, USA) statistical analysis and expert discussion, 18 alternatives were selected to provide the calibration model and two additional alternatives were used to validate the accuracy of the calibration model (i.e., alternatives 19 and 20), as detailed in Table 2.

Table 2. Attribute levels of the 20 alternatives of visits to Calligraphy Greenway.

Alternative	Total Recreational Cost (USD)	Recreational Service Quality	Recreational Activity Opportunities	Environmental Landscape Resources	Cultural Landscape Resources
1	≥32.69	Low	Participation activities only	Abundant	Abundant
2	16.35–32.69	Low	Observational activities only	Abundant	Few
3	≤16.35	Medium	Both participation and observational	Few	Abundant
4	16.35–32.69	High	Both participation and observational	Few	Abundant
5	≤16.35	Medium	Participation activities only	Abundant	Abundant
6	16.35–32.69	Low	Both participation and observational	Abundant	Abundant
7	≥32.69	Medium	Both participation and observational	Abundant	Few
8	≥32.69	Medium	Observational activities only	Abundant	Abundant
9	16.35–32.69	High	Participation activities only	Abundant	Abundant
10	≤16.35	Low	Observational activities only	Few	Abundant
11	16.35–32.69	Medium	Observational activities only	Abundant	Abundant
12	≥32.69	High	Observational activities only	Few	Few
13	≤16.35	High	Observational activities only	Abundant	Abundant
14	≤16.35	High	Both participation and observational	Abundant	Few
15	≥32.69	Low	Both participation and observational	Abundant	Abundant
16	16.35–32.69	Medium	Participation activities only	Few	Few
17	≤16.35	Low	Participation activities only	Abundant	Few
18	≥32.69	High	Participation activities only	Abundant	Abundant
19	16.35–32.69	High	Both participation and observational	Few	Few
20	≤16.35	Low	Both participation and observational	Abundant	Few

Resource from: [20]; Conducted by this study.

2.4. Study Area

The Calligraphy Greenway is in Taichung City, the second largest city in Taiwan. It has been one of the most significant attractions of Taichung. It is located in the Western district in Taichung, as shown in Figure 2. The Western district includes the National Museum of National Science, which is located at the north part of the Calligraphy Greenway and National Taiwan Museum of Fine Arts, which is located at the south part of the Calligraphy Greenway. By combining these special meaningful Taichung sights together, the Calligraphy Greenway is not only a park area in Taichung but also a designing commercial area for the tourism.

Calligraphy Greenway

Figure 2. Geographic location of the Calligraphy Greenway.

2.5. Questionnaire Design

The respondents of the questionnaire are the visitors to the Taichung Calligraphy Greenway. The rating method and the ranking method are used to investigate visitors' preferences for various alternatives. Visitors' characteristics in the survey include visitors' social background: gender, age, educational level, occupation, income and place of residence. Past recreational experiences in the survey include whether visitors have visited the Calligraphy Greenway to participate in recreational activities in the recent year, frequency of their visits, number of hours of each visit and average travel expenses per person per visit. Evaluating the preference of each alternative is conducted by asking the visitors to give their preferences for 20 alternatives in a random sequence.

This study used the rating method, in which visitors considered five attributes for each of the 20 alternatives and then scored each alternative according to their respective preferences, with 100 representing the highest preference and 1 representing absolute dislike. The questionnaire in this study was designed to realize visitors' preference levels for the "space" of the Taichung Calligraphy Greenway. The respondents surveyed in this study include not only the visitors to the Taichung Calligraphy Greenway for recreation but also the residents that went for exercises.

There are two parts of the questionnaire. The first part is the choice experiment questions. The score that the respondents gave would be used to analyze the preference and the utility of the recreational settings. For example, one of the scenarios was that the visitor would spend less than 16.35 USD and enjoy a service with a low quality recreational setting, few cultural landscape designs and abundant environmental landscape designs. The visitors would give a score to represent how much he or she liked the scenario. The second part of the questionnaire is about the socio-demographic characteristic survey. In this part, the visitors were asked to fill up their gender, age, income, occupation, education level, residence, the motivation of visiting the Calligraphy Greenway, the frequency of visiting the Calligraphy Greenway, how long on average they spent at the Calligraphy greenway and how much they spent during the visit. This study also used five Likert scale to exam satisfaction with the Calligraphy Greenway among the visitors.

The questionnaires were distributed during July to September of 2016. It was conducted by random sampling. Respondents were surveyed at the major squares and entrances of the Calligraphy Greenway (see Figure 3). The visitors received a simple explanation of the questionnaire's purpose and instructions first. The total number of the visitors that were asked to respond is 500 but only 250 visitors were willing to respond. Some respondents did not finish all the questions in this survey because of personal reasons and hence they were not regarded as valid samples. The total number of questionnaires received was 228. After removing these invalid samples, the number of valid questionnaires is 203 [23,25]. Note that invalid questionnaires were defined as those in which responses of all questions were similar or in which some questions were not responded.

Figure 3. Major squares and entrances of the Taichung Calligraphy Greenway.

2.6. Questionnaire Analysis Method

According to [26], reliability analysis measures the internal consistency of the questionnaire. Cronbach alpha is one of the value measurement in reliability analysis. The value of the Cronbach alpha depends on the items been examined. A high Cronbach alpha value indicates that the questionnaire or the survey has internal consistency. Additionally, Pearson's coefficient analysis investigates the relationship between two variables. It measures the strength of the association between the two variables. If the coefficient is positive, the two variables would increase or decrease simultaneously; otherwise, one of the variables would increase (respectively, decrease) while the other variable decreases (respectively, increase) [27].

Multilinear regression method has been widely used as a technique for decision making [27]. It investigates the relationship between independent variables and the dependent variables. The results of the multilinear regression method could indicate how the independent variables influence the dependent variable. In addition, it could be used to explain how much the independent variables could explain the dependent variables, which this study is interested in. In this study, the relative importance method is also used to evaluate the importance of attributes [20]. It uses different alternative sets to weight the importance of attributes, which could show the preference of the respondents toward these attributes [20].

3. Results and Discussion

3.1. Questionnaire Reliability Analysis

This study conducted reliability analyses on valid questionnaires. The Cronbach's α value of this questionnaire exceeds 0.7. Considering this study had 20 items, 0.7 in the Cronbach value states that the questionnaire (which is tested by the reliability test) is internally consistent.

3.2. Analysis of Descriptive Statistics

3.2.1. Analyzing Frequency of Visitors' Basic Attributes

As shown in Table 3, the data analysis for the visitor sample surveyed in this study shows a greater number of female visitors, with a total of 114, or 56.2%. 171 visitors, or 84.2% identified themselves as "unmarried." The majority of visitors, a total of 144, or 70.9%, reported a "college" level of education. A total of 89 visitors, or 43.8%, reported their occupation as "student." The second most common occupation was "service industry," with a total of 21 persons, or 10.4%. 17 visitors, or 8.4% identified themselves as "self-employed." The reason why the number of student visitors is large may be that the questionnaire was distributed from July to September of 2016, which was the summer break of students in Taiwan.

Table 3. Demographic statistics of the visitor sample in this study.

Dimension	Variable	Quantity (People)	Percentage (%)
Gender	Male	89	43.8
	Female	114	56.2
Marital Status	Unmarried	171	84.2
	Married	32	15.8
Educational Level	Middle school	2	1.0
	High school/vocational school	23	11.4
	University/trade school	144	70.9
	PhD	34	16.7
Employment Status	Compulsory military service	8	3.9
	Finance	12	5.9
	Trade/commerce	7	3.4
	Service Industry	21	10.4
	Agriculture (forestry, aquaculture, animal husbandry)	2	1.0
	Traditional manufacturing	10	4.9
	Electronics, tech, or information industry	12	5.9
	Research or educational institution	3	1.5
	Student	89	43.8
	Seeking employment	11	5.4
	Housekeeper	4	2.0
	Retired	3	1.5
	Self-employed	17	8.4
	Other	4	2.0
Motivation	Exercise	14	6.9
	Taking a walk	137	67.5
	Shopping	9	4.4
	Visiting	19	9.4
	Passing by	21	10.3
	Participating in a Calligraphy Greenway activity	3	1.5
Residence	Nearby resident	152	75.9
	Non-local (Taiwanese)	48	23.6
	Other countries	1	0.5

Note: Conducted by this study.

In the section concerning the recreational motivation for visiting the Calligraphy Greenway, a total of 137, or 67.5% of visitors reported "taking a walk" as their motivation. Because the Calligraphy Greenway is an urban green space located in the midst of a bustling urban residential area, visitors chose this location for brief periods of relaxation. Most visitors were "nearby local residents," with a total of 152 visitors, or 75.9%. It is also possible that the lack of large exhibitions or special events (e.g., Taichung Jazz Fest) might account for the low volume of non-local visitors, as the distance to visit the Calligraphy Greenway might be too far for the non-local visitors. The previous study also showed that the distance to the recreation location would have negative influence on the visitors' intension [28,29].

3.2.2. Analysis of the Descriptive Statistics for Visitors' Attributes

As shown in Table 4, the average age of visitors is 27.08 years. Visitors' average annual salary was 8993.73 USD, which is lower than Taiwan's current average annual salary of 20,383.04 USD [30], which may result from the greater number of students. In the past year, the overall average number of visits to the Calligraphy Greenway was 9.15, most of which were local visitors. This is consistent with studies for urban forests [28], in which the visitors who live closer to the recreation area would tend to visit the area more frequently.

Table 4. Analysis of the descriptive statistics for basic visitor attributes.

	Variable	Average	Standard Deviation	Rank
	Age (year)	27.08	9.84	
	Annual income (326.93 USD)	27.51	25.80	
	Number of visits to the greenway over the past year	9.15	7.44	
	Time spent at the Calligraphy Greenway (hour)	2.37	1.41	
Satisfaction	Recreational service quality	3.71	80	3
	Recreational activity opportunities	3.85	72	1
	Environmental landscape resources	3.83	75	2
	Cultural landscape resources	3.64	77	4
	Overall satisfaction	3.90	56	

Note: Conducted by this study; ranks for the level of satisfaction with the Calligraphy Greenway: 1 (very unsatisfied), 2 (unsatisfied), 3 (no opinion), 4 (satisfied), 5 (very satisfied).

As an urban green space, the Calligraphy Greenway's ensures that it can be easily accessed. The average length of a visit to the Calligraphy Greenway was 2.37 h. Statistical analysis indicates that the majority of visitors traveled to the Calligraphy Greenway for the purpose of leisure recreation. They viewed it as a place to take breaks during the workday. The visitors were satisfied with the recreational attributes. People were most satisfied with "recreational activity opportunities," assigning it an average score of 3.85 points, then "environmental landscape resources," with an average satisfaction score of 3.83 points. There is relatively lower satisfaction with design of the cultural landscape and planning.

Note that the total satisfaction is higher than the single satisfaction because we asked the respondents to give a score for each satisfaction but the four aspects that we asked did not cover all aspects that influence satisfaction.

3.3. Analysis of the Descriptive Statistics for Preferences in Terms of Attributes

As shown in Table 5, of the 20 alternative cases, Alternative 9 (with recreational cost between 16.35–32.69 USD, high quality of recreational services, availability of only participation-based recreational activities, abundant environmental landscape resources and abundant cultural landscape resources) scores the highest preference, at 86.79 points.

Table 5. An overall statistical analysis of the Calligraphy Greenway.

Alternative	Total Recreational Cost (USD)	Recreational Service Quality	Recreational Activity Opportunities (Participation, Observational)	Environmental Landscape Resources	Cultural Landscape Resources	Average	Rank
1	≥32.69	Low	Participation only	Abundant	Abundant	56.80	19
2	16.35–32.69	Low	Observational only	Abundant	Few	56.06	20
3	≤16.35	Medium	Both	Few	Abundant	68.64	8
4	16.35–32.69	High	Both	Few	Abundant	71.85	5
5	≤16.35	Medium	Participation only	Abundant	Abundant	72.64	4
6	16.35–32.69	Low	Both	Abundant	Abundant	64.30	12
7	≥32.69	Medium	Both	Abundant	Few	64.79	11
8	≥32.69	Medium	Observational only	Abundant	Abundant	68.36	9
9	16.35–32.69	High	Participation only	Abundant	Abundant	86.79	1
10	≤16.35	Low	Observational only	Few	Abundant	62.68	15
11	16.35–32.69	Medium	Observational only	Abundant	Abundant	69.67	7
12	≥32.69	High	Observational only	Few	Few	58.91	18
13	≤16.35	High	Observational only	Abundant	Abundant	77.96	3
14	≤16.35	High	Both	Abundant	Few	78.57	2
15	≥32.69	Low	Both	Abundant	Abundant	62.68	16
16	16.35–32.69	Medium	Participation only	Few	Few	63.20	13
17	≤16.35	Low	Participation only	Abundant	Few	61.16	17
18	≥32.69	High	Participation only	Abundant	Abundant	70.76	6
19	16.35–32.69	High	Both	Few	Few	63.16	14
20	≤16.35	Low	Both	Abundant	Few	64.83	10

Note: Conducted by this study; evaluation scores range from 1–100.

The second highest score is for Alternative 14 (with recreational cost of 16.35 USD or lower, high quality of recreational services, availability of both participation and observation-based activities, abundant environmental landscape resources and few cultural landscape resources), with an average score of 78.57 points. The alternative with the lowest score is Alternative 2 (with recreational cost between 16.35–32.69 USD, low quality of recreational services, availability of only observational activities, abundant environmental landscape resources and few cultural landscape resources), with an average score of 56.06 points. Of the 20 alternatives, only Alternative 9 (with recreational cost between 16.35–32.69 USD, high quality of recreational services, availability of only participation-based recreational activities, abundant environmental landscape resources and abundant cultural landscape resources) scores an average score of more than 80 points. Alternative 1 (with recreational cost of 32.69 USD or more, high quality of recreational services, availability of only participation activities, abundant environmental landscape resources and abundant cultural landscape resources), Alternative 2 (with recreational cost between 16.35–32.69 USD, low quality of recreational services, availability of only observational activities, abundant environmental landscape resources and few cultural landscape resources) and Alternative 12 (with recreational cost of 32.69 USD or more, high quality of recreational services, availability of only observational recreational activities, few environmental landscape resources and few cultural landscape resources) score less than 60 points. Other alternatives average between 60–80 points.

3.4. Analyzing Preferences for the Calligraphy Greenway by the Rating Method

3.4.1. Analyzing Preferences for the Recreational Attributes

The evaluation results using the rating method for the Calligraphy Greenway are shown in Table 6, in which the attribute utility value indicates the utility of an attribute level chosen by visitors. The total recreational cost of more than 16.35 USD has a negative attribute utility for each alternative. Total recreational cost of 16.35 USD or less has a positive attribute utility value. This shows that an increase in total recreational cost reduces the attribute utility of visitors, which is consistent with the results of previous studies [13,20]. The amount of total recreational cost affects the utility of the alternative choice. The results show that the greater the overall recreational cost for visiting the Calligraphy Greenway, the lower the visitor utility and the lower the preference.

Table 6. Evaluation results for the Calligraphy Greenway using the rating method.

Attribute	Preference Using the Rating Method (N = 203)		
	Coefficient	Attribute Utility Value	Attribute Relative Importance
Total recreational cost	−0.007		
≥32.69 USD		−0.811	22.55%
16.35–32.69 USD		−0.465	
≤16.35 USD		1.276	
Recreational service quality			
High quality		1.257	34.08%
Medium quality		0.674	
Low quality		−1.931	
Recreational activity opportunities			
Only participation activities		−0.325	11.44%
Only observational activities		−0.298	
Both participation and observational activities		0.623	
Environmental landscape resources			
Abundant		0.676	17.79%
Few		−0.676	
Cultural landscape resources			
Abundant		0.537	14.14%
Few		−0.537	
Cox & Snell R^2		0.123	
Nagelkerke R^2		0.123	
McFadden R^2		0.024	

In terms of the "recreational service quality" attribute, high and medium quality of recreational services are all positive and the attribute utility value for high quality of recreational services is greater for visitors. Low recreational service quality results in a negative attribute utility value. For visitors to the Calligraphy Greenway, a quality of service facility that is too low leads to an extremely negative recreational experience. This result shows that the higher the recreational service quality, the greater the utility value [28,31].

The results show that abundant environmental landscape resources and cultural landscape resources have a positive effect on the attribute utility values of visitors, so both environmental and cultural landscape resources influence recreational utility for visitors. The result is consistent with the results of previous studies. For example, the environmental landscape [20,32] and cultural landscape [20] have a positive influence on the visitors' preferences. The statistics show that abundant environmental landscape resources have a greater utility value than abundant cultural landscape resources. This shows that increasing environmental landscape resources has a greater impact on increasing visitors' recreational utility than increasing cultural resources. Overall, an increase in either environmental landscape resources or cultural landscape resources increases the recreational utility for visitors.

For visitors who visited the Calligraphy Greenway, the attribute with the greatest utility was the recreational service quality (34.08%), so visitors considered the recreational service quality to be the most important factor, which is in agreement with the results of previous studies [28,31]. The recreational facilities that visitors considered to be most important include walking paths, street lamps, public restrooms and trash bins. It may be inferred that the recreational service quality exerted the most direct impact on visitors' recreational experience because visitors regarded these recreational facilities as the most significant [28,31]. Bertram et al. [28] indicated that cleanliness and maintenance are important preferences for visitors. The cleanliness, for instance, is measured by the setting of garbage cans, which is regarded as recreation services in our study. In addition, Lin [20] found that recreation services are also the most significant attribute among all.

The recreational attribute that was regarded as the second most important was the cost of travel to reach the Calligraphy Greenway (22.55%). It can be inferred that, because the Calligraphy Greenway is located in the center of the city and the recreational group consists mostly of local residents, the recreational cost exerted a certain degree of influence on visitors' recreational preferences [5].

The next most important attribute is the abundance of environmental landscape resources and cultural landscape resources, the relative importance of which are 17.79% and 14.14%, respectively. In terms of the greater perceived importance of environmental landscape resources compared to cultural landscape resources, previous studies also showed that visitors were willing to spend a greater amount of money to visit areas in which nature has been more fully preserved [33]. The utility values for this attribute showed that increasing the abundance of natural environment would result in a higher utility value.

Finally, the least important attribute was the recreational activity opportunities that the Calligraphy Greenway provides (11.44%). This analysis of basic attributes shows that the majority of visitors (a total of 67.5%) visited to take walks leisurely. An analysis of these utility values for attributes shows that providing only observational activities results in a greater utility value than providing only participation activities, which further shows that visitors were not particularly interested in actively participating in the events held at the Calligraphy Greenway. This infers that opportunities for recreational activities constitute a recreational attribute that is of lesser importance. Note that the categories of recreation activities are designed especially for the Calligraphy Greenway. Most of the previous studies used positive or negative recreation activities as their categories. For example, Brey and Lehto [34] categorized recreation activities into a positive class, a nondescript class and a negative class. However, it is difficult for the visitors to distinguish between positive and negative activity categories. Therefore, this study classifies the recreational activity opportunities into observational activities and participation activities. Our study is a pioneer for the new classification for recreational activities.

3.4.2. Gender Analysis for Preferences for Recreational Attributes

Previous studies have determined the effect of gender on participation. It has also been shown that gender can influence recreational preferences [35]. Hence, this study analyzes whether gender affects the degree of preference for different forms of recreation. The visitors are divided into two groups according to gender, with a total of 114 females and 89 males (Table 7). A comparison of the rating values for preferences shows that males and females displayed similar trends in terms of the effect of attributes. Females attached a greater importance than males to the amount of total recreational cost and the availability of recreational opportunities. By contrast, males placed a greater emphasis than females on the recreational service quality, environmental landscape resources and cultural landscape resources. While these preferences are not particularly obvious, it is inferred that these results show that females attached greater importance to abstract social contact [35], interpersonal interactions and feelings than do males; and that males attached greater emphasis to concrete and practical applications of facilities and hardware.

Table 7. Comparison of the results for the preferences of male and female visitors to the Calligraphy Greenway.

Attribute	Female (N = 114)		Male (N = 89)	
	Attribute Utility Value	Relative Importance of Attribute	Attribute Utility Value	Relative Importance of Attribute
Total recreational cost				
≥32.69 USD	−0.798	22.66%	−0.838	22.81%
16.35–32.69 USD	−0.548		−0.366	
≤16.35 USD	1.346		1.204	
Recreational service quality				
High quality	1.256	34.01%	1.267	34.17%
Medium quality	0.758		0.576	
Low quality	−1.306		−1.843	
Recreational activity opportunities				
Participation activities only	−0.359	11.45%	−0.286	11.43%
Observational activities only	−0.279		−0.322	
Both participation and observational activities	0.683		0.608	
Environmental landscape resources				
Abundant	0.729	17.77%	0.610	18.17%
Few	−0.729		−0.610	
Cultural landscape resources				
Abundant	0.439	14.11%	0.661	14.14%
Few	−0.439		−0.661	
Cox & Snell R^2	0.121		0.130	
Nagelkerke R^2	0.122		0.131	
McFadden R^2	0.023		0.026	
Pearson's R	0.606 *** ($p \leq 0.0001$)		0.026 *** ($p \leq 0.0001$)	

Note: Conducted by this study; *** $p < 0.001$.

The most important attribute was the recreational service quality. The greater the recreational service quality, the higher the utility value. This value was similar for both males and females. Males attached a slightly greater importance to recreational service quality (34.17%) than do females (34.01%). The greater the total recreational cost, the lower the attribute utility. This shows that visitors had a greater preference for lower total recreational cost, which is consistent with the results of previous studies. The statistics also show that females attached greater importance to total recreational cost (22.66%) than did males (22.81%), which shows that females had higher requirements in terms of total recreational cost than did males.

There is no significant difference between males and females in terms of preferences for environmental and cultural landscape resources. Males awarded environmental landscape resources an importance of 18.17% and females awarded it an importance of 17.77%. Males awarded cultural landscape resources an importance of 14.14% and females awarded it an importance of 14.11%. Only these two attributes show no significant differences but environmental landscape resources and cultural landscape resources exerted a slightly greater influence on males. Finally, both males and females attached least importance to the opportunities of recreational activities held at the Calligraphy Greenway. Females preferred observational activities to participation activities; and men preferred participation activities.

3.4.3. Analyzing Residents' Preferences in Terms of Recreational Attributes

Questionnaire visitors were divided in terms of their place of residence. 153 visitors live in Taichung city and 50 are non-residents (also called non-locals), which is consistent with the results of previous studies [28].

The results for preference evaluation is shown in Table 8, in which locals attach greater importance to total recreational cost and recreational activity opportunities than did non-locals. Non-locals attach greater importance to environmental landscape resources and cultural landscape resources. Both groups, to some extent, attached a similar level of importance to the recreational service quality.

Table 8. Comparison of the results for the preferences of local and non-local visitors to the Calligraphy Greenway.

Attribute	Locals (N = 153)		Non-locals (N = 50)	
	Attribute Utility Value	Relative Importance of Attribute	Attribute Utility Value	Relative Importance of Attribute
Total recreational cost				
≥32.69 USD	−0.835	22.61%	−0.750	22.37%
16.35–32.69 USD	−0.482		−0.412	
≤16.35 USD	1.317		1.162	
Recreational service quality				
High quality	1.201	34.08%	1.437	34.08%
Medium quality	0.640		0.780	
Low quality	−1.841		−2.217	
Recreation activity opportunities				
Participation activities only	−0.346	11.46%	−0.277	11.39%
Observation activities only	−0.314		−0.265	
Both participation and observation activities	0.660		0.542	
Environmental landscape resources				
Abundant	0.609	17.75%	0.885	17.93%
Few	−0.609		−0.885	
Cultural landscape resources				
Abundant	0.532	14.10%	0.562	14.23%
Few	−0.532		−0.562	
Cox & Snell R^2		0.117		0.146
Nagelkerke R^2		0.118		0.147
McFadden R^2		0.023		0.028
Pearson's R		0.606 *** ($p \leq 0.001$)		0.726 *** ($p \leq 0.000$)

Note: Conducted by this study; *** $p < 0.001$.

The statistics for total recreational cost show that the relative importance of this attribute was slightly greater for locals (22.61%) than for non-locals (22.37%), which shows that locals might attach slightly more importance to total recreational cost than did non-locals. On the reason for which locals attached greater importance to total recreational cost, it is inferred that the most visits to the Calligraphy Greenway aimed to take a leisurely walk and the visitors did not deliberately plan their trips. Previous studies also showed that locals were less flexible in terms of their willingness to change the environment than were outsiders [28], so there was less overall willingness among locals to spend a greater sum of money on recreation.

Environmental landscape resources might have a greater importance for non-locals (17.93%) than for locals (17.75%). In terms of cultural landscape resources, non-locals awarded this attribute an importance of 14.23% and locals awarded it an importance of 14.10%, which is an insignificant difference. However, environmental landscape resources exerted a slightly greater influence on non-locals than did cultural landscape resources. It is inferred that, because non-locals traveled to Taichung from outside areas, their first impression of the recreational area was visual, so non-locals were likely to have a more significant reaction to landscape installations, so they awarded greater levels of importance to these two attributes.

Finally, locals and non-locals regarded the recreational activity opportunities provided by the Calligraphy Greenway as least important. This factor might have a greater level of importance for locals (11.46%) than for non-locals (11.39%). A detailed comparison shows that, for both locals and non-locals, only providing observational activities resulted in a greater utility than only providing participation activities, which shows that visitors preferred that the Calligraphy Greenway provides observational activities to participation activities.

3.5. A Multilinear Regression Analysis of the Results

This study investigated which aspects of visitor background affected the frequency, the length of stay and the overall satisfaction of visitors to the Calligraphy Greenway. These three items constitute the independent variables for a multilinear regression analysis and the analysis results are given in Table 9. "Number of visits to the Calligraphy Greenway in the past year" and "age" show a clear positive correlation (β = 0.186 **), indicating elder people visited the Calligraphy Greenway more frequently. This result is in agreement with those for previous studies [28,35,36]. Previous studies also noted that visitors to urban forests or green spaces tended to be elder. For example, the result in [35] showed that the elder would tend to be living in walkable neighborhoods. The walking friendly environment, such as urban forests or city greenways, would encourage elder visitors to visit the area [35,36]. The advantageous location of the urban forest and green space plays a role in this trend.

Table 9. Multilinear regression analysis of the data.

Dimension		Number of Visits to the Calligraphy Greenway Over the Past Year	Time of Stay at the Calligraphy Greenway	Overall Satisfaction with the Calligraphy Greenway
Constant		13.190 *	1.436	1.460 ***
Gender		−0.437	−0.391	−0.062
Age		0.186 **	−0.012	−0.002
Marital status		0.413	−0.317	−0.110
Educational level		−3.547 ***	0.183 *	−0.027
Annual income		0.050 *	−0.003	-6.185×10^{-5}
Residence location		−3.123 ***	0.086	0.077
Number of visits to the Calligraphy Greenway over the past year		—	0.015 ***	0.007
Length of visit to the Calligraphy Greenway		1.313 ***	—	0.023
Amount of money spent at the Calligraphy Greenway		0.000	6.253×10^{-5}	-2.678×10^{-5}
Satisfaction with the recreational attributes	Recreational service quality	−1.028	0.032	0.177 ***
	Recreational activity opportunities	1.217	0.072	0.095 *
	Environmental landscape resources	0.688	−0.319 *	0.117 *
	Cultural landscape resources	−1.185	−0.127	0.284 ***
Overall satisfaction		1.630	0.268	—
R^2		0.336	0.178	0.555
Adjusted R^2		0.336	0.112	0.519
Significance		0.000 ***	0.001 ***	0.000 ***
Durbin-Watson Test		1.805	2.260	2.134

Note: Conducted by this study; * $p < 0.05$, ** $p < 0.01$, *** $p < 0.001$.

There is a negative correlation with "educational level" ($\beta = -3.547$ ***). Those with higher levels of education were less likely to come to Calligraphy greenway.

There is a negative correlation with "place of residence" ($\beta = -3.123$ ***). A previous study by Bertram and Larondelleb [5] showed that the distance between home and a recreation area impacts the frequency with which people participate in recreation. An analysis of the results shows that visitors who reside closer to the Calligraphy Greenway visited the site more frequently. It is inferred that because they live closer to the park, reaching the recreation area is easier, so visitors were more willing to participate in recreational activities in this place. These results are consistent with those of previous studies [37]. There is a positive correlation with "time of stay at the Calligraphy Greenway" ($\beta = 1.313$ ***), so the more frequently visitors make visits, the longer are their stays, because a greater frequency of visits indicates that a visitor already relies heavily on recreation at the Calligraphy Greenway, so he or she prioritizes this area when considering recreational options.

There is a positive correlation between "time of stay at the Calligraphy Greenway" and "educational level" ($\beta = 0.183$ *), so the higher a visitor's level of education, the longer the stay at the Calligraphy Greenway, which is consistent with the results of previous studies [4,37]. The time of stay has a positive correlation ($\beta = 0.015$ ***) with "number of visits to the Calligraphy Greenway over the past year," so the more regularly a person visited the Calligraphy Greenway, the more likely they were to linger at the Calligraphy Greenway. This may be because those who visited the Calligraphy Greenway more frequently lived in close vicinity to the park, so they were willing to spend longer time at the Calligraphy Greenway. The time of stay has a negative correlation ($\beta = -0.319$ *) with "satisfaction with environmental landscape resources," so those with lower levels of satisfaction with the environmental landscape resources spent longer time at the Calligraphy Greenway. This may be due to the fact that the environmental landscape of the Calligraphy Greenway was not a primary motivation for people who spent longer time at the Calligraphy Greenway. Motivational reasons for this group tend to be activities (such as picnicking, soccer, or flying kites). Previous studies also showed that visitors' recreational experience varies for different modes of recreation [38]. Visitors to the Calligraphy Greenway also had varying recreational experiences and levels of satisfaction, depending on the different activities in which they participated.

The "overall satisfaction with the Calligraphy Greenway" has a positive correlation with "satisfaction with the recreation and service quality of the Calligraphy Greenway" ($\beta = 0.177$ ***) and "satisfaction with cultural landscape resources" ($\beta = 0.284$ ***). Each of these four items is used to evaluate satisfaction with the Calligraphy Greenway, so each item has a high positive correlation with the overall satisfaction.

3.6. Correlation Analysis of the Results

This study analyzed the interactive relationship between pairs of visitors' social background. Multilinear Regression Analysis was used to analyze the frequency with which visitors visited the Calligraphy Greenway for recreation, their stay time and their satisfaction levels. A Pearson correlation analysis is used to analyze the correlation between variable pairs, as shown in Table 10.

Table 10. Correlation analysis between variable pairs.

	Gender	Age	Marital Status	Income	Residential Location	Number of Visits	Time of Stay	Total Costs	Service Quality	Activity Opportunities	Environmental Landscape	Cultural Landscape	Overall Satisfaction
Gender	1												
Age	0.175 *	1											
Marital status	0.054	0.629 ***	1										
Income	0.209 **	0.566 ***	0.490 ***	1									
Residential location	0.016	0.006	−0.042	0.002	1								
Number of visits	−0.029	0.332 ***	0.260 ***	0.268 ***	−0.302 ***	1							
Length of stay	−0.177 *	−0.071	−0.081	−0.038	−0.059	0.249 ***	1						
Total costs	0.067	−0.053	−0.023	0.035	−0.012	−0.077	0.023	1					
Service quality	−0.052	−0.065	−0.012	−0.089	0.069	−0.004	−0.024	−0.063	1				
Activity opportunities	−0.094	−0.105	0.014	0.001	−0.126	0.127	0.064	−0.109	0.459 ***	1			
Environmental landscape	0.031	−0.038	0.027	−0.016	−0.031	0.037	−0.114	0.019	0.518 ***	0.389 ***	1		
Cultural landscape	0.031	−0.200 **	−0.093	−0.135	−0.059	−0.091	−0.063	−0.087	0.350 ***	0.390 ***	0.434 ***	1	
Overall satisfaction	−0.092	−0.183 **	−0.117	−0.140 *	0.007	0.056	0.059	−0.137	0.544 ***	0.475 ***	0.496 ***	0.599 ***	1

Note: Conducted by this study; * $p < 0.05$, ** $p < 0.01$, *** $p < 0.001$.

Age and marital status show a positive correlation with gender ($r = 0.629$). A greater proportion of male visitors to the Calligraphy Greenway were married than were female visitors. These results correspond with the marital status statistics of Taichung City. The total population of married males in Taichung City is greater than the total population of married females [30,39]. Age and income show a positive correlation ($r = 0.566$) for visitors to the Calligraphy Greenway, so older people had a higher income level. Marital status and income show a positive correlation ($r = 0.490$). Visitors who were married had higher incomes than visitors who were not married, which corresponds with marital status and income data from Taichung City Government [39].

"Satisfaction with the recreational service quality" and "satisfaction with the recreational activity opportunities" show a positive correlation ($r = 0.459$), so visitors with a higher degree of satisfaction with the recreational service quality also have a higher degree of satisfaction with the recreational activity opportunities. "Satisfaction with the recreational service quality" has a positive correlation with the "satisfaction with environmental landscape resources" ($r = 0.518$), so visitors who had a higher degree of satisfaction with the recreational service quality also had a higher degree of satisfaction with the environmental landscape resources. "Satisfaction with the recreational service quality" has a positive correlation with the "satisfaction with cultural landscape resources" ($r = 0.518$), so visitors who had a higher degree of satisfaction with the recreational service quality also had a higher degree of satisfaction with the cultural landscape resources. "Satisfaction with the recreational service quality" had a positive correlation with the "overall satisfaction" ($r = 0.544$), so visitors who exhibited a greater degree of satisfaction with the recreational service quality had a greater overall satisfaction.

"Satisfaction with the recreational activity opportunities" and the "overall satisfaction" shows a positive correlation ($r = 0.475$), so visitors who had a greater degree of satisfaction with the recreational activity opportunities had a greater degree of the overall satisfaction. "Satisfaction with environmental landscape resources" and "satisfaction with cultural landscape resources" shows a positive correlation ($r = 0.434$), so visitors who had a greater degree of satisfaction with environmental landscape resources had a greater degree of satisfaction with cultural landscape resources. "Satisfaction with environmental landscape resources" and "the overall satisfaction" shows a positive correlation ($r = 0.496$), so visitors who had a greater degree of satisfaction with environmental landscape resources had a greater degree of overall satisfaction. "Satisfaction with the cultural landscape resources" and "the overall satisfaction" show a positive correlation ($r = 0.599$), so the greater a visitor's degree of satisfaction with the cultural landscape resources, the greater their overall satisfaction.

This study has the following two hypotheses: (1) the visitors to the Calligraphy Greenway had different preference levels for recreational attributes and (2) the visitors with different socioeconomic backgrounds had different preference levels for recreational attributes. The results for hypothesis (1) show that the visitors had the highest preference for service quality of recreational facilities but attached the lowest importance to cultural landscape resources. The results for hypothesis (2) show that the female visitors attached greater importance to the total recreational cost than the male; local visitors were more concerned about the total recreational cost than non-local visitors; and all visitors attached the greatest importance to the service quality of recreational facilities.

4. Conclusions and Recommendations

This study analyzes citizens' preferences for recreational facilities and determines ways in which facilities and landscape of the Calligraphy Greenway in Taichung in Taiwan can be improved and provides a valuable reference for future city greenway planning. The goal of this study is to analyze (1) the effect of visitors' preference levels to the Calligraphy Greenway on different recreational attributes and (2) the effect of preference levels of visitors with different socioeconomic backgrounds on different recreational attributes. This study has used a CE method to evaluate the preferences of visitors to the Calligraphy Greenway in terms of five attributes: total recreational cost, recreational service quality, recreational activity opportunities, environmental landscape resources and cultural landscape resources.

Correlation analysis, relative importance method and multilinear regression analysis are applied in this study. The visitors' levels of preference for various recreational attributes were explored and the resulting data was used to improve the current recreational quality. The significance of attributes was ranked from high to low, in terms of recreational service quality, total recreational cost, environmental landscape resources, cultural landscape resources and recreational activity opportunities. That is, the recreational service quality is the most important attribute for the respondents, whereas the cultural landscape resources are the less one. Results for attribute levels show that as total recreational cost increased, visitors' utility decreased; as the recreational service quality increases, visitors' utility increased. The results for recreational activity opportunities show that observational activities had a high utility for visitors and that both environmental and cultural landscape resources had a positive utility for visitors. The results of this study show that the visitors to the Calligraphy Greenway tended to prefer the recreational types with low consumption. Karanikola et al. [40] suggested that more attention should be paid to the groups with a low income and limited recreation, because they accounted for about 60% of all visitors. They also mentioned that visitors had higher preference levels (expectation) for recreational facilities. Therefore, their results are consistent with ours.

This study also uses multilinear regression analysis to investigate the effect of preference levels of visitors with different socioeconomic backgrounds on different recreational attributes. The statistical data was divided into male and female groups for comparison. Females attached greater importance to total recreational cost and recreational activity opportunities than did males; and males attached greater importance to recreational facility quality, environmental landscape resources and cultural landscape resources than did females. In addition, the data was divided into local and non-local residents. Locals attached greater importance to total recreational cost and recreational activity opportunities than did non-locals; and non-locals attached greater importance to environmental and cultural landscape resources than did locals. Both groups attached a similar degree of importance to the recreational service quality. Although the result does not show statistical significance, the trend still shows a slight difference due to the effects of different social demographic characteristics.

In terms of the patterns for "number of visits to Calligraphy Greenway over the past year," the result shows that the older the visitor, the greater the frequency of visits to the Calligraphy Greenway; and the longer the stay at the Calligraphy Greenway is, the more frequently a visitor comes to Calligraphy Greenway for recreation. The visitors with lower educational levels visited more frequently. The nearer local residents lived, the more frequently did they visit. In the research to "time of stay at the Calligraphy Greenway," the visitors with higher levels of education stayed for longer and the greater the frequency of visits, the longer is the stay. The analysis result of "overall satisfaction with the Calligraphy Greenway" shows that, the greater the degree of satisfaction with each item, the greater the overall satisfaction. In addition, the results of this study show that most visitors went for a walk in the Taichung Calligraphy Greenway and the majority of visitors were Taichung residents. The result shows that most of the visitors to an urban forest/greenway would be those who lived closer to it [40,41]. The function of urban green space and forests tended to provide an area for relaxation, going for a walk and doing exercises. Hence, from recreational attributes of visitors, it was observed that most visitors were the residents in the neighborhood. The results of this study show that most visitors visited there ten times per year on average, which is more frequent than found in the study conducted by Sreethran [41]. Most of the visitors to the Calligraphy Greenway paid more attention to environmental landscape resources than cultural landscape resources. This result is consistent with the results of previous studies for preferences on urban forests or green space, in which most visitors preferred to enjoy the feeling of being close to the nature [40,41].

Based on these results, recommendations for the improvement of the recreational quality for the Calligraphy Greenway were made as follows:

1. The government should improve the design for the cultural landscape and increase the quality of cultural landscape resources and recreational value, in order to encourage visitors to visit. It was found that the greenway area has more resources that can be applied and managed, as compared with other recreational areas. Hence, the authorities should make use of these resources to promote integration of recreation and cultural landscape resources.
2. The authorities should make more efforts to improve integration of recreation and cultural landscape resources. For instance, from the viewpoint of tour guide, more information boards should be set up to provide visitors the historical backgrounds of the Calligraphy Greenway. Interpretation boards for art installations of humanistic landscape should also be set up to enhance the benefits of integrating recreation and cultural landscape.
3. The government should improve the Calligraphy Greenway's facility to provide visitors with better recreational service quality, in order to encourage visitors to participate in recreation and to increase their willingness to visit. From the results of this study, the attribute with the greatest utility was the service quality of recreational facilities. That is, if recreational service quality could be promoted, the most benefit could be obtained. Therefore, this study suggests that the government should first conduct an overall inspection on recreational service facilities, including the uneven parts of walkways and unpaired fences in public areas. The first priority is to ensure safe travel routes.
4. The government should create customized plans for different visitor groups to accommodate varying preferences for recreational attributes, in order to improve recreational experiences for various recreational groups.

Author Contributions: Investigation, C.C.; Supervision and Funding Acquisition, W.-Y.L.; Writing-original draft, W.-Y.L. and C.C.

Funding: This research was funded in part by the Ministry of Science and Technology, Taiwan, and the Council of Agricultural, Taiwan.

Acknowledgments: The authors thank the anonymous referees for comments that improved the content as well as the presentation of this paper.

Conflicts of Interest: The authors declare no conflict of interest.

References

1. European Commission. Environment: Green Infrastructure. 2016. Available online: http://ec.europa.eu/environment/nature/ecosystems/index_en.htm (accessed on 20 May 2018).
2. Gobster, P.H.; Westphal, L.M. The human dimensions of urban greenways: Planning for recreation and related experiences. *Landsc. Urban Plan.* **2004**, *68*, 147–165. [CrossRef]
3. Weber, S.; Boley, B.B.; Palardy, N.; Gaither, C.J. The impact of urban greenways on residential concerns: Findings from the Atlanta BeltLine Trail. *Landsc. Urban Plan.* **2017**, *167*, 147–156. [CrossRef]

4. Japelja, A.; Mavsarb, R.; Hodgesc, D.; Kovačd, M.; Juvančiče, L. Latent preferences of residents regarding an urban forest recreation setting in Ljubljana, Slovenia. *For. Policy Econ.* **2015**, *71*, 71–79. [CrossRef]
5. Bertram, C.; Larondelleb, N. Going to the woods is going home: Recreational benefits of a larger urban forest site—A travel cost analysis for Berlin, Germany. *Ecol. Econ.* **2017**, *132*, 255–263. [CrossRef]
6. Lin, Y.-X. An Examination of Urban Greenway Development in Taichung City. Master's Thesis, Feng Chia University, Taichung, Taiwan, 2008.
7. Taichung City Government. Taichung City Geographic Information System Map. 2018. Available online: http://gismap.taichung.gov.tw/address/index.cfm (accessed on 20 May 2018). (In Chinese)
8. Taichung City Government Tourism Office. Taichung Travel Net: Calligraphy Greenway. Taichung City: Taichung City Government Tourism Office. 2018. Available online: https://travel.taichung.gov.tw/en-us/Attractions/Intro/1050/Calligraphy-Greenway (accessed on 20 May 2018).
9. FAO, Food and Agriculture Organization of the United Nation. Global Ecological Zones (Second Edition) Map. 2018. Available online: http://ref.data.fao.org/map?entryId=2fb209d0-fd34-4e5e-a3d8-a13c241eb61b&tab=about (accessed on 20 May 2018).
10. Wu, K.P. A Study of Visitors' Satisfaction with Urban Parkway Facilities—A Case Study on Ching-Kuo Parkway in Taichung City. Master's Thesis, Feng Chia University, Taichung, Taiwan, 2002.
11. FIABCI World Prix d' Excellence Awards, Past Winners, Winners of Prix d' Excellence Awards 2010. Available online: http://fiabciprix.com/2010-winners/ (accessed on 20 May 2018).
12. Dehez, J.; Lyser, S. Combining multivariate analysis and cost analysis in outdoor recreation planning. *J. Outdoor Recreat. Tour.* **2014**, *7–8*, 75–88. [CrossRef]
13. Campbell, D.; Vedel, S.E.; Thorsen, B.J.; Jacobsen, B.J. Heterogeneity in the WTP for recreational access: Distributional aspects. *J. Environ. Plan. Manag.* **2014**, *57*, 1200–1219.
14. Tu, G.; Abildtrupa, J.; Garciaa, S. Preferences for urban green spaces and peri-urban forests: An analysis of stated residential choices. *For. Policy Econ.* **2016**, *70*, 56–66. [CrossRef]
15. Louviere, J.J.; Hensher, D.A.; Swait, J. *Stated Choice Methods: Analysis and Application*; Cambridge University Press: Cambridge, UK, 2000.
16. Lancaster, K.J. A new approach to consumer theory. *J. Political Econ.* **1966**, *2*, 132–157. [CrossRef]
17. McFadden, D. Conditional logit analysis of qualitative choice behavior. In *Frontiers in Econometrics*; Zarembka, P., Ed.; Academic Press: New York, NY, USA, 1973; pp. 105–142.
18. Boxall, P.C.; Adamowicz, W.L. Understanding heterogeneous preferences in random utility models: A latent class approach. *Environ. Resour. Econ.* **2002**, *23*, 421–446. [CrossRef]
19. Darmon, R.Y.; Rouzies, D. Internal validity of conjoint analysis under alternative measurement produces. *J. Bus. Res.* **1999**, *46*, 67–81. [CrossRef]
20. Lin, Y.-J. Study on Recreation Site Choice Behavior: Application of Stated Preference Model. *J. Outdoor Recreat. Res.* **2000**, *13*, 63–86.
21. Lin, P.-Y.; Liaw, S.-C. Applying the Fuzzy Delphi Method to analyze the greenway functions of Lover River in Kaohsiung city. *J. Exp. For. Natl. Taiwan Univ.* **2008**, *22*, 89–106.
22. Yuan, Y.-L.; Lue, C.-C. Understanding the relationships between recreation experience and perception to management actions in forestry settings using qualitative approach. *Q. J. Chin. For.* **2007**, *40*, 55–68.
23. Bai, H.-Z. Exploring the Charms of Taichung Calligraphy Greenway. Master's Thesis, Tunghai University, Taichung, Taiwan, 2014.
24. Sardana, K.; Bergstrom, J.C.; Bowker, J.M. Valuing setting-based recreation for selected visitors to national forests in the southern United States. *J. Environ. Manag.* **2016**, *183*, 972–979. [CrossRef] [PubMed]
25. Wann, J.-W.; Yang, Y.-C.; Huang, W.-S.; Lin, Y.-F. An empirical analysis of consumer's willingness to pay for attributes of domestic banana: A study in metropolitan areas in Taiwan. *J. Agric. For.* **2013**, *62*, 249–265.
26. Babbie, E. *The Practice of Social Research*, 14th ed.; Change Learning Press: Boston, MA, USA, 2016.
27. Hair, J.F., Jr.; Black, W.C.; Babin, B.J.; Anderson, R.E. *Multivariate Data Analysis*, 7th ed.; Pearson Education Press: London, UK, 2010.
28. Bertram, C.; Meyerhoff, J.; Rehdanz, K.; Wüstemann, H. Differences in the recreational value of urban parks between weekdays and weekends: A discrete choice analysis. *Landsc. Urban Plan.* **2017**, *159*, 5–14. [CrossRef]
29. Schipperijn, J.; Ekholm, O.; Stigsdotter, U.K.; Toftager, M.; Bentsen, P.; Kamper-Jørgensen, F.; Randrup, T.B. Factors influencing the use of green space: Results from a Danish national representative survey. *Landsc. Urban Plan.* **2010**, *95*, 130–137. [CrossRef]

30. Executive Yuan Budget, Accounting and Statistics Office. Taiwan. 2018. Available online: https://www.dgbas.gov.tw/ct.asp?xItem=33338&ctNode=3099&mp=1 (accessed on 20 May 2018).
31. Andkjær, S.; Arvidsen, J. Places for active outdoor recreation—A scoping review. *J. Outdoor Recreat. Tour.* **2015**, *12*, 25–46. [CrossRef]
32. Whiting, J.W.; Larson, L.R.; Green, G.T.; Kralowec, C. Outdoor recreation motivation and site preferences across diverse racial/ethnic groups: A case study of Georgia state parks. *J. Outdoor Recreat. Tour.* **2017**, *18*, 10–21. [CrossRef]
33. Sulaiman, F.C.; Hasan, R.; Jamaluddin, E.R. The mature trees in recreation areas and its role in enhancing quality of life. *Soc. Behav. Sci.* **2016**, *234*, 289–298. [CrossRef]
34. Brey, E.T.; Lehto, X. The relationship between daily and vacation activities. *Ann. Tour. Res.* **2007**, *34*, 160–180. [CrossRef]
35. Ghania, F.; Rachelea, J.N.; Washingtonc, S.; Turrella, G. Gender and age differences in walking for transport and recreation: Are the relationships the same in all neighborhoods? *Prev. Med. Rep.* **2016**, *4*, 75–80. [CrossRef] [PubMed]
36. Kerr, J.; Emond, J.A.; Badland, H.; Reis, R.; Sarmiento, O.; Carlson, J.; Sallis, J.F.; Cerin, E.; Cain, K.; Conway, T.; et al. Perceived neighborhood environmental attributes associated with walking and cycling for transport among adult residents of 17 cities in 12 countries: The IPEN study. *Environ. Health Perspect.* **2016**, *124*, 290–298. [CrossRef] [PubMed]
37. Lee, K.H.; Schuett, M.A. Exploring spatial variations in the relationships between residents' recreation demand and associated factors: A case study in Texas. *Appl. Geogr.* **2014**, *53*, 213–222. [CrossRef]
38. Prayaga, P. Estimating the value of beach recreation for locals in the Great Barrier Reef Marine Park, Australia. *Econ. Anal. Policy* **2017**, *53*, 9–18. [CrossRef]
39. Budget, Accounting and Statistics Office, Taichung City Government. *Taichung City Region 2011–2015: Comparison of Marital Status Ratio among Males and Females. Taichung City Regional Income Differences*; Budget, Accounting and Statistics Office, Taichung City Government: Taichung, Taiwan, 2016. Available online: http://www.dbas.taichung.gov.tw (accessed on 20 May 2018).
40. Karanikola, P.; Panagopoulos, T.; Tampakis, S. Weekend visitors' views and perceptions at an urban national forest park of Cyprus during summertime. *J. Outdoor Recreat. Tour.* **2017**, *17*, 112–121. [CrossRef]
41. Sreethran, M. Exploring the urban use, preference and behaviors among the residents of Kuala Lumper: Malaysia. *Urban For. Urban Green.* **2017**, *25*, 85–93. [CrossRef]

© 2018 by the authors. Licensee MDPI, Basel, Switzerland. This article is an open access article distributed under the terms and conditions of the Creative Commons Attribution (CC BY) license (http://creativecommons.org/licenses/by/4.0/).

Article

Mutual Influences of Urban Microclimate and Urban Trees: An Investigation of Phenology and Cooling Capacity

Celina H. Stanley [1,2,*], Carola Helletsgruber [2] and Angela Hof [2]

[1] Department of Landscape Change and Management, Leibniz Institute of Ecological Urban and Regional Development, 01217 Dresden, Germany
[2] Department of Geography and Geology, University of Salzburg, 5020 Salzburg, Austria
* Correspondence: C.Stanley@ioer.de; Tel.: +49-351-467-9269

Received: 31 May 2019; Accepted: 21 June 2019; Published: 26 June 2019

Abstract: This paper presents an empirical study on urban tree growth and regulating ecosystem services along an urban heat island (UHI) intensity gradient. The UHI effect on the length of the growing season and the association of cooling and shading with species, age, and size of trees was studied in Salzburg, Austria. Results show that areas with a low UHI intensity differed from areas with a medium or high UHI intensity significantly in three points: their bud break began later, the leaf discoloration took longer, and the growing season was shorter. After leaves have developed, trees cool the surface throughout the whole growing season by casting shadows. On average, the surfaces in the crown shade were 12.2 °C cooler than those in the sun. The tree characteristics had different effects on the cooling performance. In addition to tree height and trunk circumference, age was especially closely related to surface cooling. If a tree's cooling capacity is to be estimated, tree age is the most suitable measure, also with respect to its assessment effort. Practitioners are advised to consider the different UHI intensities when maintaining or enhancing public greenery. The cooling capacity of tall, old trees is needed especially in areas with a high UHI intensity. In the future, species differences should be examined to determine the best adapted species for the different UHI intensities. The present results can be the basis for modeling future mutual influences of microclimate and urban trees.

Keywords: growing season; bud break; surface temperature; urban heat island; urban microclimate; urban trees

1. Introduction

Cities are the most important living space of humans, at least as measured by the proportion of the population living in them. By 2030, global urban population is projected to increase to 70%. This growing urbanization is causing huge changes in the urban environment [1]. One result of these transformations is the urban heat island (UHI) effect, which is characterized by higher air and surface temperatures compared to the rural environment. Factors contributing to this effect include, for example, the coverage of natural surfaces with material that absorbs the solar radiation more strongly and the industrial heat output. The increase in temperature has enormous negative consequences for the urban population and lowers their quality of life as air and water quality decrease, as well as the thermal comfort. A particular risk comes from the increasing and prolonged heatwaves that lead to a higher number of fatalities [2]. The projected increase in the average annual temperatures suggests an intensification in the UHI effect [3].

However, urban green spaces can mitigate the UHI effect. In contrast to sealed surfaces which have high runoff and low evaporation rates, plants remove energy from their ambient air in the form of heat and thus evaporate the water they store. Through evapotranspiration, the plants not only cool themselves but also the air temperature in the immediate vicinity [4]. Furthermore, urban trees provide indirect temperature cooling by shading the surface. The crown blocks the solar radiation and the surface heats up less. Therefore, due to less radiation, the air temperature under the crown heats up much less as well [5].

The extent of the cooling effects depends on various environmental factors and plant characteristics [4]. Many studies deal with the positive benefits of large green spaces, such as city parks, whereas urban trees are rarely in focus. Yet even individual trees show a cooling effect [6]. One study identified correlations between air temperature cooling and leaf color, foliage density, leaf thickness, and structure of the leaves [7]. Another study about surface cooling showed that with an increasing leaf area index (LAI), the asphalt temperature decreases, with this effect being independent of tree species; however, there was no corresponding effect on a grassy surface [8]. Mao et al. studied two streets, one lined with deciduous trees and the other lined with coniferous trees [9]; due to a higher evapotranspiration capacity of the deciduous trees, the temperatures on this street were lower. Differences in cooling performance occur even between different tree species. These differences are related to the trees' characteristics, such as height or crown shape [6]. Gillner and colleagues [10] compared six species of trees for their capacity to increase relative humidity and to reduce surface and air temperatures. There were significant differences in the microclimatic impact of the species. *Corylus colurna* L. and *Tilia cordata* Mill. provided particularly high cooling, whereas *Ulmus × hollandica* 'Lobel' had the lowest cooling values. The authors used leaf area density (LAD) as a tree characteristic with which they associated surface cooling. Across all tree species, there were strong positive correlations between surface cooling and LAD [10].

Though trees can mitigate UHI, not all species are adapted to it. Urban and street trees only reach on average about 25%–50% of their potential age range [11]. This is partly due to the wrong choice of tree species [11] and illustrates the need to follow up with the effects of the site conditions on the trees.

Temperature is one of the most important factors affecting the phenological development of trees. Besides bud break, flowering or leaf fall is also determined by temperature. Temperature-induced changes in phenological development were detected in different tree species and in different study areas. A study carried out in England with more than 200 species showed, for example, that tree blossom began four days earlier when the average annual temperature rose by one degree Celsius [12]. In addition to the average annual temperature, researchers of the Landesanstalt für Umwelt, Messungen und Naturschutz Baden-Württemberg (LUBW) found another temperature-related climate parameter with the cumulative sum of the average daytime temperatures, which has a connection with the occurrence of phenological phases [13]. Furthermore, an analysis of the International Phenological Gardens of Europe showed a clear gradient in bud break between the warm countries of southern Europe and the cold north [12]. Additionally, Chmielewski and Rötzer [12] found that spring temperatures decisively determine the timing of leaf emergence. The higher the air temperatures were between February and April, the earlier the buds break. This was also the case in so-called "extreme years" when the emergence started particularly early or late according to the temperature. They found that a temperature rise of one degree Celsius was associated with a growing season which began 6.7 days earlier in Europe [12]. Thus, it becomes obvious that warming due to climate change and the higher temperatures in the city generated by the UHI compared to the surrounding area must have an impact on the timing of bud break. Zipper and colleagues [14] studied the relationship between bud break and the location of the tree. There was a significantly longer (8.0–10.5 days) growing season in urban areas compared to rural areas. The length of the growing season in parks was only about 5 days longer than in rural areas. This can be explained by the park cool island effect. Despite the inner-city location of the parks, the UHI effect is less pronounced than in the built-up city due to the temperature reduction of the vegetation [14].

Assuming that the temperature, as key driver, determines the start of bud break, duration of leaf unfolding and the growing season, and that differences between impervious cover and urban green space are found, the question arises as to what extent UHI intensity is different within the built-up city and has an influence on the bud break and growing season of urban trees. In the present study, mutual effects between UHI and urban trees were investigated. On the one hand, it was determined how different UHI intensities affect the phenological phases (bud break and leaf discoloration) as well as the length of the entire growing season of urban trees. On the other hand, we analyzed to what extent urban trees reduce the surface temperature and to which tree characteristics this cooling performance is related. By combining these two aspects, recommendations for applied urban planning are made.

2. Materials and Methods

2.1. Study Sites and Objects

The study was conducted in the City of Salzburg, Austria (coordinates: 47°47′ N, 13°00′ E; altitude: 436 m). According to the Köppen climate classification, the climate in Salzburg is a marine west coast (Cfb) climate. The daily mean temperature varies between −0.8 °C in January and 18.6 °C in July. Due to the location at the northern rim of the Alps, the amount of precipitation is rather high with an amount of 1336 mm/year, mainly in the summer months [15].

In order to investigate the influence of the varying UHI intensities on bud break, first the warmer and cooler areas of the city had to be identified. For this, the number of summer days was used as a climatic parameter. Summer days are defined as all those days with a daily maximum temperature of at least 25 °C [16]. The information was obtained from a layer of the Zentralanstalt für Meteorologie und Geodynamik (ZAMG) SISSI project. The layer maps the average number of summer days per year for the period from 1971 to 2000 with a spatial resolution of 100 × 100 m. The maps were obtained from a microscale urban climate model which, among other things, takes building structure into account [17]. First, the layer had to be reduced to the urban areas with an impervious cover greater than 30%. Otherwise, the green areas, especially due to the city mountains in Salzburg, would have distorted the classification. Subsequently, a UHI intensity gradient was generated. With the help of "natural breaks" classification, the urban area was classified into three UHI-intensity categories. As the name suggests, this classification searches the values for natural breaks and ensures that the categories themselves are as homogeneous as possible while contrasting each other [18]. The least number of summer days occurs in category 1 areas, with up to 55.8 summer days per year. Accordingly, these areas have a low UHI intensity. In category 2 areas, there are 55.9–64.8 summer days per year, which is why these areas have a medium UHI intensity. In category 3 areas, there are more than 64.8 summer days per year. Accordingly, these areas have a high UHI intensity. The locations of the observed tree species were selected within these three categories, and the study sites are listed in Table 1. The selection of tree species included several considerations. In addition to the number of individuals, the Citree database played a major role. This database is a supporting tool for choosing the optimal trees for urban areas by providing information about more than 390 tree and shrub species, including site characteristics, for example [19]. By analyzing the tree register of Salzburg (a map showing all public trees), the most common species of the publicly owned and managed tree stock were selected. The higher the number of individuals, the higher the effect of the species on the urban climate. To ensure that the selected tree species were adapted to the future urban climate, the species drought tolerance, hardiness, heat tolerance, and late frost tolerance were analyzed by using the Citree database. Using the tree register, eight study sites were identified.

Table 1. Overview of the study sites with urban heat island (UHI) intensity/category, tree species, and number of trees.

Study Sites	UHI Intensity/Category	Tree Species	Number of Trees
Berchtesgadner Straße	low/1	Acer platanoides	4
Max-Reinhardt-Platz	low/1	Tilia cordata	8
Hofhaymer Allee	medium/2	Tilia cordata	6
Otto-Holzbauer-Straße	medium/2	Acer platanoides	8
Erzabt-Klotz-Straße	high/3	Tilia cordata	3
Franz-Josef-Straße	high/3	Aesculus × carnea	6
Guggenmoosstraße	high/3	Corylus colurna	6
Südtiroler Platz	high/3	Acer platanoides	5

2.2. Phenological Monitoring

For the phenological monitoring in spring, we used the well-established method presented by Wesolowski and Rowinski [20]. They developed a scale of point values from 0 to 2 for assessing the development status of a leaf bud. For each observation day, ten randomly selected apical buds in the upper, south-exposed part of the crown are evaluated and their sum is calculated [20]. The monitoring starts when all buds are closed and thus evaluated as having zero points. As soon as all ten leaves are completely developed and each scores two points, the monitoring is finished. To record the evaluation of the buds, an observation sheet was specially designed. The phenological monitoring according to Wesolowski and Rowinski's instructions was carried out every three days from 21 March 2018 to 5 May 2018.

To test this, further phenological monitoring was conducted during the autumn of 2018. Following and adapting the methodology described in the teaching and learning materials of the European Union (EU)'s COMENIUS Project named BEAGLE (Biodiversity Education and Awareness to Grow a Living Environment) [21], four scores were used:

(0) No leaf discoloration and/or leaf fall
(1) Beginning leaf discoloration and/or leaf fall (<50% of leaves)
(2) Pronounced leaf discoloration and/or leaf fall (≥50% of leaves)
(3) Complete leaf discoloration and/or leaf fall

The phenological monitoring in the autumn was done once in a week starting on September 13, 2018 and ending on November 27, 2018. By using the beginning and end of leaf discoloration, we calculated the length of leaf discoloration.

2.3. Tree Physiognomy

For all observation trees, the height, trunk circumference at breast height, and leaf area index (LAI) were measured. Using these data, the tree age, crown area, and crown volume were further calculated. The tree height was measured using a Leica DISTOTM D810 Touch (Leica Geosystems, Heerbrugg, Switzerland). To determine the LAI, an LAI-2200C Plant Canopy Analyzer from LI-COR (Lincoln, NE, USA) was used. Four LAI measurements at breast height were made per tree along the four cardinal directions using a 90° cap. By averaging the four values, differences in the crown density were balanced. The measured values were then edited in the FV2200 software from LI-COR (2.1.1, Lincoln, NE, USA). Among other things, the crown shape was specified, which was previously extracted from photos following the methodology presented in the device manual. After adaptation of the crown shape, the LAI was interpreted as foliage density for single tree measurements [22]. Thus, crown area and crown volume were calculated. Tree age was estimated using the trunk circumference. Based on earlier surveys of growth rates, tables exist for the species-specific annual growth in centimeters. If the current trunk circumference is divided by this factor, then the approximate age of the tree is obtained [23].

2.4. Microclimatic Measurements

The climatic parameter used was surface cooling. The microclimate was measured using the difference of the surface temperatures between the crown-shaded area and the full sun-exposed area (compare [10]). The measurements were carried out using an Infrared Radiometer, Model MI-220, from Apogee Instruments Inc. (Logan, UT, USA). The device was fixed on a tripod to make sure that all measurements were carried out from the same height (0.55 m) and angle (45°). To calculate the cooling effect, the surface temperature in the shade of each tree crown was measured and once per study site on a full sun-exposed reference area. For all temperature measurements, the same height and angle were used. The reference areas were as close as possible to the studied trees and had the same surface type. Because of the changing sun position, it was important to measure the surface temperature of the tree crown-shaded area as centrally as possible.

To minimize the variations of surface temperatures linked to the changing atmospheric conditions during a day and between different days, all measurements were run through on one day (July 24, 2018) between 13:30 and 15:30 under cloudless conditions.

2.5. Statistical Analysis

One-way ANOVAs were performed to assess whether the beginning of bud break and length of leaf unfolding, the onset and length of leaf discoloration, and the length of growing season differed between the three UHI intensities. In the case of significant results, Tukey post-hoc tests were conducted to test all pairwise comparisons. The correlations between surface cooling and tree characteristics (tree height, trunk circumference, age, LAI, crown area, and crown volume) were assessed using Pearson's r. For all statistical data processes, we used R version 3.4.4 (R Core Team, Vienna, Austria). The level of significance was set at $p < 0.05$.

3. Results

3.1. Effect of UHI Intensity on Tree Physiognomy

In order to assess the extent to which three different UHI intensities influence the length of the growing season and, ultimately, tree physiognomy, the results of the phenological monitoring in spring and autumn were compared with respect to the three defined categories.

Using data from the phenological monitoring in spring, a significant difference was observed between the three UHI intensities with respect to the beginning of the bud break, $F(2, 51) = 15.4$, $p < 0.001$. On average, the leaves of the trees began to develop earlier in areas with a high UHI intensity compared to areas with a low UHI intensity (Figure 1A). Significant differences were found between categories 1 and 2 ($p = 0.001$) and between categories 1 and 3 ($p < 0.001$) using Tukey post-hoc tests. On average, the bud break of the trees in categories 2 and 3 began on April 7 (SD = 2 days) and on April 5 (SD = 5 days), respectively. However, this difference was not significant ($p = 0.363$). In contrast, the leaves in category 1 began to develop on average on 13 April (SD = 4 days).

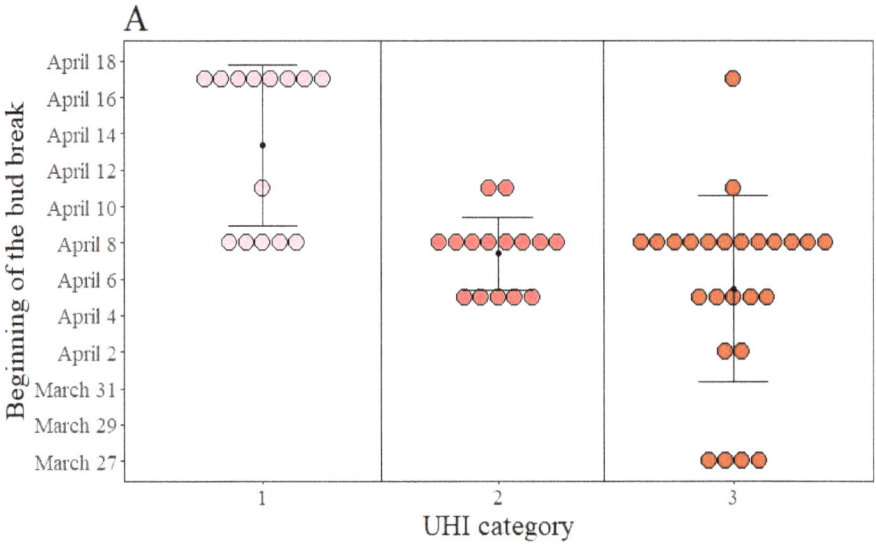

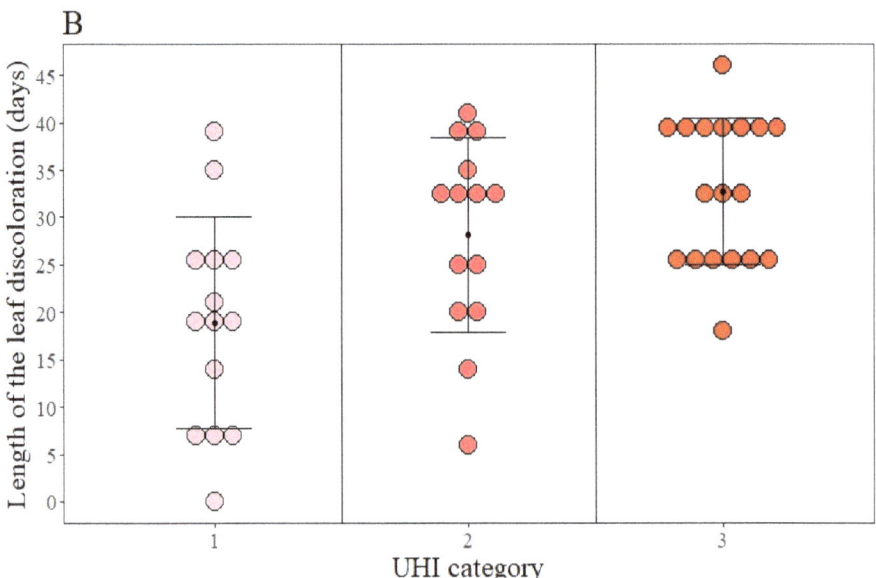

Figure 1. *Cont.*

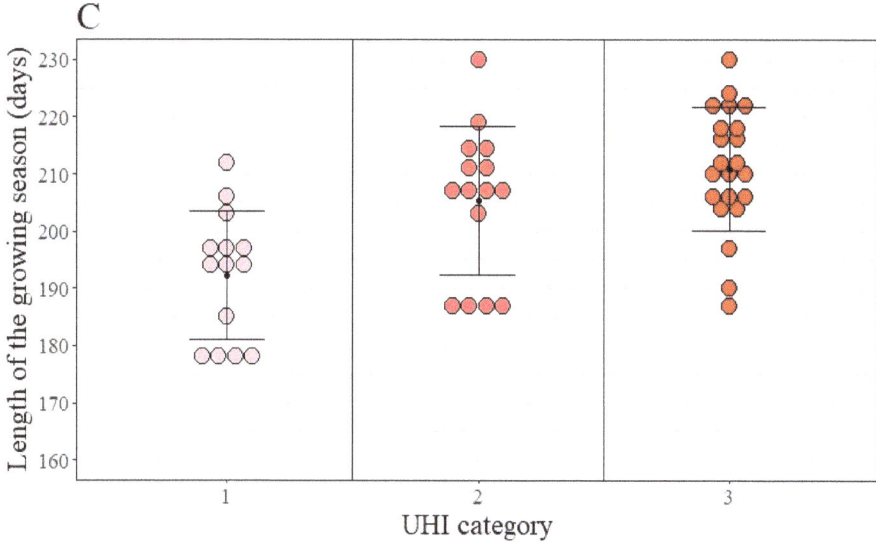

Figure 1. (**A**) The beginning of the bud break, (**B**) the length of the leaf discoloration, and (**C**) the length of the growing season of the investigated trees, separated by UHI intensity. The figure additionally shows the means and standard deviations.

The average length of the leaf unfolding varied between 11 days (SD = 3 days) in areas with a low UHI intensity and 13 days in areas with a medium (SD = 3) or high (SD = 7 days) UHI intensity. However, this difference was not significant, $F(2, 51) = 1.0, p = 0.379$.

The phenological monitoring in autumn provided information on the relationship between the onset and the length of leaf discoloration with UHI intensity. The first tree started to discolor on September 13 and the last on October 18. On average, the trees in areas with a medium UHI intensity started to discolor earliest, namely, on September 30 (SD = 7 days). In comparison, the trees in the warmest urban areas began to discolor on average on October 3 (SD = 6 days) and the trees in the coolest areas only on October 5 (SD = 8 days). However, the differences between the three categories were not significant, $F(2, 45) = 2.1, p = 0.134$.

In contrast, there was a significant difference between the three UHI intensities with respect to the length of leaf discoloration, $F(2, 45) = 8.8, p < 0.001$. On average, complete discoloration took 19 days (SD = 11 days) in areas with a low UHI intensity, 29 days (SD = 11 days) in areas with a medium UHI intensity, and 34 days (SD = 8 days) in areas with a high UHI intensity (Figure 1B). Using Tukey post-hoc tests, we found that category 1 was significantly different from both category 2 ($p = 0.019$) and category 3 ($p < 0.001$), but there was no significant difference between categories 2 and 3 ($p = 0.466$).

Finally, the two monitoring periods (spring and autumn) were combined to determine the length of the growing season (Figure 1C). The period between the beginning of bud break and the completion of leaf discoloration differed significantly among the three UHI intensities, $F(2, 50) = 12.7, p < 0.001$. Using Tukey post-hoc tests, significant differences were found between categories 1 and 2 ($p = 0.014$) and between categories 1 and 3 ($p < 0.001$) but not between categories 2 and 3 ($p = 0.166$). The average length of the growing season was 192 days (SD = 11 days) in category 1, 205 days (SD = 13 days) in category 2, and 213 days (SD = 12 days) in category 3.

3.2. Estimation of Surface Cooling Using Tree Characteristics

The average surface cooling of the examined trees was 12.2 °C (SD = 3.8 °C). The values ranged from at least 6.4 °C up to a maximum cooling of 22.4 °C.

For a more detailed investigation of surface cooling, all trees were aggregated regardless of tree species or UHI intensity. Table 2 gives an overview of the investigated tree characteristics that were associated with surface cooling.

Table 2. Overview of tree characteristics with mean values, standard deviations, minima, and maxima.

Tree Characteristic	Mean (SD)	Min	Max
Tree height (m)	10.9 (2.2)	7.0	16.0
Trunk circumference (cm)	91.2 (30.3)	43.0	142.0
Age (years)	33.4 (11.1)	17.0	56.0
LAI (m^2/m^2)	1.5 (0.7)	0.5	3.1
Crown area (m^2)	32.6 (13.2)	8.9	63.0
Crown volume (m^3)	129 (79.4)	19.1	338.0

From these six criteria, the one that best predicts surface cooling can be identified. For this purpose, the correlations were determined between surface cooling and tree height, trunk circumference, age, LAI, crown area, and crown volume (Table 3). Surface cooling turned out to be most strongly related to age, followed by trunk circumference and tree height. The tree characteristics themselves were also strongly interrelated. All the correlations listed in Table 3 were significant ($p < 0.05$).

Table 3. Overview of correlations among the tree characteristics and surface cooling.

Tree characteristics	1	2	3	4	5	6	7
1 Surface cooling	-						
2 Tree height	0.51 **	-					
3 Trunk circumference	0.55 ***	0.85 ***	-				
4 Age	0.61 ***	0.77 ***	0.93 ***	-			
5 LAI	−0.34 *	−0.45 **	−0.49 **	−0.41 *	-		
6 Crown area	0.47 **	0.72 ***	0.81 ***	0.72 ***	−0.49 **	-	
7 Crown volume	0.45 **	0.77 ***	0.83 ***	0.72***	−0.46 **	0.95 ***	-

* The correlation is significant at the $p < 0.05$ level, ** The correlation is significant at the $p < 0.01$ level, *** The correlation is significant at the $p < 0.001$ level.

4. Discussion

In this study, the mutual influence of urban trees and UHI intensity was examined. On the one hand, the effects of UHI intensity on tree phenology were investigated. On the other hand, the effectiveness of trees in cooling the surface was considered with regard to their characteristics. Significant effects were found for both research questions which are discussed separately below.

4.1. Effect of UHI Intensity on Tree Phenology

The phenological monitoring in spring and autumn showed significant differences between the UHI intensities with respect to the beginning of the bud break, the length of the leaf discoloration, and the length of the growing season. The trees in cooler areas began later with their bud break, their leaves were completely discolored after a shorter time in the autumn, and their entire growing season was significantly shorter. It can thus be deduced that as temperatures increase, the bud break is accelerated and the growing season is prolonged. These findings are consistent with those of other studies. It was frequently stated that the elevated temperatures are a major reason for the increasingly early beginning of the bud break [12,20,24]. An increase in temperature of one degree Celsius causes phenological processes around the world to begin four to six days earlier, which is why the results of phenological observations are also useful indicators of the progress of climate change [24]. Above all, earlier or greater warming in the spring promotes an earlier beginning of the bud break [25]. Some of the studies on the influence of temperature on the phenology of trees are not confined to urban areas but also include forests [12]. As little was known about the trees' response specifically to the elevated temperatures in the city compared to the surrounding area, Zipper and colleagues [14] investigated this issue with a longitudinal study over three years. They were able to prove that UHI is reflected in the phenology of the trees. Compared to the surrounding area, the growing season of trees in the city was on average five days longer. Measurements in urban parks showed a weakened effect regarding the extended length of the growing season, which was explained by the park cool island effect [14]. The results of the present study substantiate the relationship between temperature rise and premature beginning of bud break as well as extended length of the growing season from another point of view. It is not the general temperature increase that is investigated, but rather inner-urban temperature variations. Even within the urban area, UHI intensity or temperature differences occur. These temperature differences have an impact on tree phenology regardless of what causes them, whether it be the degree of impervious cover, the density of buildings, or wind blocks. However, our data show no continuous progression between the three UHI intensities regarding the three parameters, namely, beginning of the bud break, length of the leaf discoloration, and length of the growing season. Instead, there were significant differences between trees in areas of categories 1 and 2 as well as categories 1 and 3. Therefore, it is reasonable to assume that there is a threshold value in the temperature at which, for example, the beginning of the bud break is forced. Regarding the UHI intensities, the biggest phenological differences occur between categories 1 and 2. If the UHI intensity increases even further, from that point on, for example, no significantly earlier beginning of the bud break is expected. The realization that the city is a small-scale mosaic of UHI intensities and that this has an impact on the phenology of the trees turns out to be extremely relevant for urban planning practices. Our findings show that it is worth taking into account the local UHI intensity when choosing tree species to maximize the benefits, as urban trees may grow larger and provide more ecosystem services if the urban forest is strategically planted [26]. In cooler areas, species with a long growing season should therefore be planted to benefit from the regulating ecosystem services of the trees as early as possible and well into the autumn. There are already studies on these species-specific reactions to temperature and their phenological characteristics which will be crucially important in the future [24,27]. This makes it possible to find species which can adapt to rising temperatures and at the same time make the city more resilient against them [28]. In addition, more research attention should be paid to the second research question in the present study, that is, how significant is the cooling effect of the trees and which tree characteristics are associated with it.

4.2. Estimation of Surface Cooling Using Tree Characteristics

All investigated tree characteristics correlated moderately to strongly with surface cooling. Only the LAI correlated negatively with both surface cooling and the other tree characteristics. The negative correlation between the LAI and surface cooling contradicts the results of other studies [8,10]. A full discussion of these findings is beyond the scope of the present paper. Instead, the focus is on the other tree characteristics that were positively related to surface cooling. Nevertheless, it is worth noting that, through its frequent use, the LAI is established as a proven tool for predicting the climate impact of trees [8]. Despite all of this, the LAI also has weaknesses. A comparison of different studies that determined the LAI in a direct and indirect way showed an underestimation of the LAI when using measuring instruments for indirect estimation. This underestimation varied between 25% to 50% [29]. In addition, the effort needed to collect the LAI can be determined from the method description, which is another reason to find an alternative measure. Therefore, a comparison of the correlations of different tree characteristics with surface cooling should indicate which other parameters, and ones that are possibly easier to assess, are closely related to the cooling performance of a particular tree.

Two possible parameters, tree height and crown volume, have already shown positive correlations with climatic cooling. With increasing tree size and crown, the trees blocked more radiation [30]. Likewise, the density of the crown affected the amount of blocked radiation and surface cooling [10]. This ultimately resulted in a difference between the studied tree species in their potential to cool the surface. However, Gillner and colleagues also pointed out that the species had different mean heights and were therefore somewhat limited in their cooling capacity [10]. Therefore, in the present study, the data measured and collected for the investigated species were combined in order to make statements solely about the tree characteristics.

Age showed the greatest correlation with surface cooling. In addition, the age was strongly related to the trunk circumference. This is due to the calculation method of a tree's age, which is a combination of a species-specific factor and the trunk circumference [23]. That is why the calculated tree age is only an approximation of the real tree age and is, among other things, influenced by geographic location. With regard to the assessment methods of the individual tree characteristics and the correlations of these with surface cooling, the tree age seems particularly suitable for predicting the surface cooling capacity of a tree. Since age, unless known, is calculated from the trunk circumference, it is unaffected by trimming, in contrast to the other tree characteristics. Compared to the mere circumference of the trunk, age has the advantage of taking into account species-specific growth rates, with only a minor additional effort.

Nevertheless, this does not mean that age is responsible for the cooling capacity of a tree. There exist other external factors like water availability that influence the cooling capacity. Due to a lack of water, the evaporation capacity of the trees is reduced and thus also their cooling capacity [10].

5. Conclusions

The present study has shown that, on the one hand, UHI is not uniform throughout the urban area and that the different intensities affect the phenological development of trees. When planting trees with the intention of positively influencing the urban climate, it is recommended that planners consider the city as a small-scale UHI mosaic. On the other hand, differences in tree characteristics affect the cooling of the UHI. Based on the interrelation of UHI and city trees, practical recommendations for action are given. In areas with a more pronounced UHI, it is even more important to plant species that grow tall and form a large crown to mitigate UHI in the best possible way. Small ornamental trees are not an adequate substitute from a climatological point of view. Further research on species-specific reactions to the different UHI intensities is needed. Thus, for cooler or warmer areas, the optimal species can be selected. In cooler areas, these are trees that develop their leaves even at lower temperatures and provide cooling as early as possible. In warmer areas, the species must be particularly resistant to heat, so as not to prematurely throw off their leaves in hot summers. For this purpose, comprehensive studies over several years would be useful. Based on the measured developments, future scenarios could then be created using modeling. The number of studies on the influence of green infrastructure on the temperature has increased rapidly in recent years [31–33]. For such modeling, basic work such as the present one is important because it provides input parameters. As has been shown recently, tree characteristics are also influenced by climate change and the standardized parameters collected in the past are no longer up-to-date. Among other factors, rising temperatures and the associated extension of the growing season are cited as the cause of the increased wood growth and the simultaneously reduced wood density; as a result, the estimated amount of carbon stored by trees is lower than assumed [34]. In line with this, the question arises as to how climate change will affect urban trees and their ecosystem services in future, since the trees not only mitigate the UHI effect, for example by cooling the surface, but are also themselves influenced by the UHI. This influence will intensify in the future and more research and monitoring is needed to understand this relationship. In exploring the interrelations of microclimate and urban trees, future developments must also be considered. This is the best way to ensure that the urban population can fully and most efficiently benefit from the ecosystem services of urban trees in the long term.

Author Contributions: Conceptualization, C.H.S. and A.H.; Data curation, C.H.S.; Formal analysis, C.H.S.; Funding acquisition, A.H.; Investigation, C.H.S. and C.H.; Methodology, C.H.S. and A.H.; Project administration, A.H.; Supervision, A.H.; Validation, C.H.S.; Visualization, C.H.S.; Writing—original draft, C.H.S., C.H., and A.H.; Writing—review and editing, C.H.S., C.H. and A.H.

Funding: This work was supported by the Sparkling Science research program of the Federal Ministry of Science, Research and Economy (BMWFW), Austria, project "Urban trees as climate messengers", grant number SPA 06/005. We are grateful for financial support provided by the Open Access Publication Fund of the University of Salzburg.

Acknowledgments: Open Access Funding by the University of Salzburg. We thank Christian Stadler, as well as Michael Heinl (Stadt:Gärten Salzburg), Günther Nowotny, Matthias Marbach, Laurenz Fiala, and Sten Gillner for administrative and technical support.

Conflicts of Interest: The authors declare no conflict of interest. The funders had no role in the design of the study; in the collection, analyses, or interpretation of data; in the writing of the manuscript; or in the decision to publish the results.

References

1. Nikodinoska, N.; Paletto, A.; Pastorella, F.; Granvik, M.; Franzese, P.P. Assessing, valuing and mapping ecosystem servides at city level: The case of Uppsala (Sweden). *Ecol. Model.* **2018**, *368*, 411–424. [CrossRef]
2. Leal Filho, W.; Echevarria Icaza, L.; Neht, A.; Klavins, M.; Morgan, E.A. Coping with the impacts of urban heat islands. A literature based study on understanding urban heat vulnerability and the need for resilience in cities in a global climate change context. *J. Clean. Prod.* **2018**, *171*, 1140–1149. [CrossRef]
3. Gill, S.E.; Handley, J.F.; Ennos, A.R.; Pauleit, S. Adapting cities for climate change: The role of the green infrastructure. *Built Environ.* **2007**, *33*, 115–133. [CrossRef]
4. Gago, E.J.; Roldan, J.; Pacheco-Torres, R.; Ordóñez, J. The city and urban heat islands: A review of strategies to mitigate adverse effects. *Renew. Sustain. Energy Rev.* **2013**, *25*, 749–758. [CrossRef]
5. Loughner, C.P.; Allen, D.J.; Zhang, D.-L.; Pickering, K.E.; Dickerson, R.R.; Landry, L. Roles of Urban Tree Canopy and Buildings in Urban Heat Island Effects: Parameterization and Preliminary Results. *J. Appl. Meteorol. Climatol.* **2012**, *51*, 1775–1793. [CrossRef]
6. Bowler, D.E.; Buyung-Ali, L.; Knight, T.M.; Pullin, A.S. Urban greening to cool towns and cities: A systematic review of the empirical evidence. *Landsc. Urban Plan.* **2010**, *97*, 147–155. [CrossRef]
7. Lin, B.-S.; Lin, Y.-J. Cooling Effects of Shade Trees with Different Characteristics in a Subtropical Urban Park. *HortScience* **2010**, *45*, 83–86. [CrossRef]
8. Rahman, M.A.; Moser, A.; Gold, A.; Rötzer, T.; Pauleit, S. Vertical air temperature gradients under the shade of two contrasting urban tree species during different types of summer days. *Sci. Total Environ.* **2018**, *633*, 100–111. [CrossRef]
9. Mao, L.-S.; Gao, Y.; Sun, W.-Q. Influences of street tree systems on summer micro-climate and noise attenuation in Nanjing City, China. *Arboric. J.* **1993**, *17*, 239–251. [CrossRef]
10. Gillner, S.; Vogt, J.; Tharang, A.; Dettmann, S.; Roloff, A. Role of street trees in mitigating effects of heat and drought at highly sealed urban sites. *Landsc. Urban Plan.* **2015**, *143*, 33–42. [CrossRef]
11. Roloff, A. *Bäume in der Stadt*; Ulmer Verlag: Stuttgart, Germany, 2013.
12. Chmielewski, F.-M.; Rötzer, T. Response of tree phenology to climate change across Europe. *Agric. For. Meteorol.* **2001**, *108*, 101–112. [CrossRef]
13. Holz, I.; Franzaring, J.; Böcker, R.; Fangmeier, A. *Eintrittsdaten phänologischer Phasen und ihre Beziehung zu Witterung und Klima*; LUBW: Karlsruhe, Germany, 2011.
14. Zipper, S.C.; Schatz, J.; Singh, A.; Kucharik, C.J.; Townsend, P.A.; Loheide II, S.P. Urban heat island impacts on plant phenology: Intra-urban variability and response to land cover. *Environ. Res. Lett.* **2016**, *11*, 1–12. [CrossRef]
15. Salzburg, Austria Köppen Classification: Marine West Coast Climate. Available online: https://www.weatherbase.com/weather/weather-summary.php3?s=5111&cityname=Salzburg,+Austria (accessed on 12 June 2019).
16. Zentralanstalt für Meteorologie und Geodynamik (ZAMG) Klimadaten von Österreich 1971–2000. Available online: http://www.zamg.ac.at/fix/klima/oe71-00/klima2000/klimadaten_oesterreich_1971_fr (accessed on 12 June 2019).
17. Zentralanstalt für Meteorologie und Geodynamik (ZAMG) Stadtklima Zukunft. Available online: https://www.zamg.ac.at/cms/de/klima/informationsportal-klimawandel/daten-download/stadtklima-zukunft (accessed on 12 June 2019).

18. ESRI Inc. (Environmental Systems Research Institute) Datenklassifikationsmethoden. Available online: https://pro.arcgis.com/de/pro-app/help/mapping/layer-properties/data-classification-methods.htm (accessed on 12 June 2019).
19. Vogt, J.; Gillner, S.; Hofmann, M.; Tharang, A.; Dettmann, S.; Gerstenberg, T.; Schmidt, C.; Gebauer, H.; Van de Riet, K.; Berger, U.; et al. Citree: A database supporting tree selection for urban areas in temperate climate. *Landsc. Urban Plan.* **2017**, *157*, 14–25. [CrossRef]
20. Wesolowski, T.; Rowinski, P. Timing of bud burst and tree-leaf development in a multispecies temperate forest. *For. Ecol. Manag.* **2006**, *237*, 387–393. [CrossRef]
21. Batorczak, A. The Beagle Project. In *Biodiversity in Education for Sustainable Development – Reflection on School-Research Cooperation*; Ulbrich, K., Settele, J., Benedict, F.F., Eds.; Pensoft Publishers: Sofia, Bulgaria, 2010; pp. 53–55.
22. LI-COR, Inc. *LAI-2200C Plant Canopy Analyzer Instruction Manual*; LI-COR: Lincoln, NE, USA, 2013.
23. Plietzsch, A. Die Lebensdauer von Bäumen und Möglichkeiten zur Altersbestimmung. In *Jahrbuch der Baumpflege 2009*; Dujesiefken, D., Ed.; Haymarket Media Gmbh: Braunschweig, Germany, 2008; pp. 172–188.
24. Flynn, D.F.B.; Wolkovich, E.M. Temperature and photoperiod drive spring phenology across all species in a temperate forest community. *New Phytol.* **2018**, *219*, 1353–1362. [CrossRef] [PubMed]
25. Fu, Y.H.; Campioli, M.; Deckmyn, G.; Janssens, I.A. The Impact of Winter and Spring Temperatures on Temperate Tree Budburst Dates: Results from an Experimental Climate Manipulation. *PLoS ONE* **2012**, *7*, e47324. [CrossRef] [PubMed]
26. Endreny, T.A. Strategically growing the urban forest will improve our world. *Nat. Commun.* **2018**, *9*, 1160. [CrossRef] [PubMed]
27. Caffarra, A.; Donnelly, A. The ecological significance of phenology in four different tree species: Effects of light and temperature in bud burst. *Int. J. Biometeorol.* **2011**, *55*, 711–721. [CrossRef]
28. McPherson, E.G.; Berry, A.M.; van Doorn, N.S. Performance testing to identify climate-ready trees. *Urban For. Urban Green* **2018**, *29*, 28–39. [CrossRef]
29. Bréda, N.J.J. Ground-based measurements of leaf area index: A review of methods, instruments and current controversies. *J. Exp. Bot.* **2003**, *54*, 2403–2417. [CrossRef] [PubMed]
30. Gómez-Muñoz, V.M.; Porta-Gándara, M.A.; Fernández, J.L. Effect of tree shades in urban planning in hot-arid climatic regions. *Landsc. Urban Plan.* **2010**, *94*, 149–157. [CrossRef]
31. Ambrosini, D.; Galli, G.; Mancini, B.; Nardi, I.; Sfarra, S. Evaluating Mitigation Effects of Urban Heat Islands in a Historical Small Center with the ENVI-Met®Climate Model. *Sustainability* **2014**, *6*, 7013–7029. [CrossRef]
32. Zuvela-Aloise, M.; Koch, R.; Buchholz, S.; Früh, B. Modelling the potential of green and blue infrastructure to reduce urban heat load in the city of Vienna. *Clim. Chang.* **2016**, *135*, 425–438. [CrossRef]
33. Salata, F.; Golasi, I.; Petitti, D.; de Lieto Vollaro, E.; Coppi, M.; de Lieto Vollaro, A. Relating microclimate, human thermal comfort and health during heat waves: An analysis of heat island mitigation strategies through a case study in an urban outdoor environment. *Sustain. Cities Soc.* **2017**, *30*, 79–96. [CrossRef]
34. Pretzsch, H.; Biber, P.; Schütze, G.; Kemmerer, J.; Uhl, E. Wood density reduced while wood volume growth accelerated in Central European forests since 1870. *For. Ecol. Manag.* **2018**, *429*, 589–616. [CrossRef]

© 2019 by the authors. Licensee MDPI, Basel, Switzerland. This article is an open access article distributed under the terms and conditions of the Creative Commons Attribution (CC BY) license (http://creativecommons.org/licenses/by/4.0/).

Article

The Influence of Individual-Specific Plant Parameters and Species Composition on the Allergenic Potential of Urban Green Spaces

Susanne Jochner-Oette [1,*], Theresa Stitz [1], Johanna Jetschni [1] and Paloma Cariñanos [2]

1. Physical Geography/Landscape Ecology and Sustainable Ecosystem Development, Catholic University of Eichstätt-Ingolstadt, Eichstätt 85072, Germany; Theresa.Stitz@ku.de (T.S.); johanna.jetschni@ku.de (J.J.)
2. Department of Botany, Faculty of Pharmacy, Campus de Cartuja, University of Granada, Granada 18071, Spain; palomacg@ugr.es
* Correspondence: susanne.jochner@ku.de; Tel.: +49-8421-93-21742

Received: 3 May 2018; Accepted: 19 May 2018; Published: 23 May 2018

Abstract: Green planning focusses on specific site requirements such as temperature tolerance or aesthetics as crucial criteria in the choice of plants. The allergenicity of plants, however, is often neglected. Cariñanos et al. (2014; Landscape and Urban Planning, 123: 134–144) developed the Urban Green Zone Allergenicity Index (I_{UGZA}) that considers a variety of plant characteristics to calculate the allergenic potential of urban green spaces. Based on this index, we calculated an index for the individual-specific allergenic potential (I_{ISA}) that accounts for a varying foliage volume by accurate measurements of crown heights and surface areas occupied by each tree and only included mature individuals. The studied park, located in Eichstätt, Germany, has an area of 2.2 ha and consists of 231 trees. We investigated the influence of species composition using six planting scenarios and analysed the relationship between allergenic potential and species diversity using Shannon index. Only a small number of trees was female and therefore characterised as non-allergenic, 9% of the trees were classified as sources of main local allergens. The allergenic potential of the park based on literature values for crown height and surface was I_{UGZA} = 0.173. Applying our own measurements resulted in I_{ISA} = 0.018. The scenarios indicated that replacing trees considered as sources of main local allergens has the strongest impact on the park's allergenic potential. The I_{UGZA} offers an easy way to assess the allergenic potential of a park by the use of a few calculations. The I_{ISA} reduces the high influence of the foliage volume but there are constraints in practicability and in speed of the analysis. Although our study revealed that a greater biodiversity was not necessarily linked to lower index values, urban green planning should focus on biodiversity for ameliorating the allergenic potential of parks.

Keywords: urban parks; landscape planning; allergenic potential; ecosystem disservices

---

1. Introduction

Allergic diseases are considered as important human health issues as they substantially restrict many allergic people [1,2]. In Germany, almost 20% of the adult population suffers from at least one type of allergy [1] and 30% of the adult population and 50% of the adolescent generation show sensitisations [2]. Since 15% of the population are affected by hay fever [3], pollen allergy poses a major risk for humans.

In light of these rising numbers of allergic diseases, the consideration of allergy-friendliness in urban green planning seems to be essential [1,4]. Actually, pollination is regarded as an ecosystem service provided by green infrastructure [5]. When related to pollen allergies, however, it can also be considered as an *ecosystem disservice* [6] representing a conflict to other ecosystem services such as recreation and health benefits as green spaces regulate climate and improve air quality [5].

Urban green planning mainly focusses on specific site requirements (e.g., temperature tolerance, pest resistance, tolerance to pruning) and aesthetics as crucial criteria in the choice of plants [7]. The allergenic potential of plants, however, is often neglected [1,4,8]. Still, more research is required regarding the use of allergenic plants in green spaces and the development of allergy-friendly green areas in urban environments. Some attempts have been made to formulate planning recommendations to reduce negative health impacts caused by plants producing allergenic pollen [9–11]. For example, it was recommended to develop gardens with only female plants or with a great diversity of non-allergenic plants [10]. Microbial diversity was found to positively affect the human immune system by reducing certain allergenic and respiratory diseases (e.g., reviewed by [12]). A study conducted in eastern Finland showed that atopic individuals were associated with lower environmental biodiversity in the surroundings of their homes and significantly lower generic diversity of gammaproteobacteria on their skin [13]. Since there is a link between richness of macroorganisms and the associated microbial biodiversity [12], a greater biodiversity of plants in general may reduce the risk of allergy sensitisations. In addition, an increased biodiversity linked to a reduction of traditional species with a high allergenic capacity leads to lower concentrations of monospecific pollen [14].

The allergenic potential of plants was assessed by [10] who developed the Ogren Plant Allergy Scale (OPALS). This scale considers different plant-specific criteria based on studies about similarities between allergenic and non-allergenic plants. Characteristics include pollen weight and size, pollen moisture, flower fragrance as well as the position of male flowers of monoecious species. The scale categorises species at ten levels with 1 being non-allergenic and 10 being highly allergenic. Another attempt to categorise plants according to their allergological characteristics was made by [15] who developed the allergen index (A.I.). This index is based on information on the plant's life cycle, the length of its phenanthesic period, the presence of phenomena of cross reactivity and species abundance [15,16]. In addition, [17] developed the so-called Urban Green Zone Allergenicity Index (I_{UGZA}) to calculate the allergenic potential of urban green spaces. This index compares an existent green space with a hypothetical space with maximum allergenicity. The index is based on the following assumptions: Plants with a higher crown volume emit higher pollen quantities [18], anemophilous trees produce more pollen than other trees [19] and pollen release is directly proportional to the number of individuals belonging to one species.

In this study, the current allergenic potential of the trees and shrubs in an urban park (Hofgarten, located in Eichstätt, Germany) was examined using I_{UGZA}.

Based on this index, we developed I_{ISA}, an index for the individual-specific allergenic potential of urban green spaces. Therefore, we measured the height and the surface covered by each individual tree or shrub and only included mature individuals already emitting pollen. In addition, we investigated the influence of species composition on the allergenic potential of the park using six different planting scenarios and analysed the relationship between allergenic potential and species diversity using Shannon index. Thus, our main aims were to develop and evaluate management tools (indices and planting scenarios) in order to formulate recommendations for urban planning. Although this study only includes a single park, our methods are also applicable for any park of any size and therefore illustrative for the practicability of the management tools presented in our study.

2. Materials and Methods

2.1. Study Area

The studied park Hofgarten (48°53′20.54″ N, 11°11′17.78″ E, 385 m a.s.l.) is located in the city of Eichstätt, Bavaria, Germany (Figure 1a,b). The climate is temperate with an average annual temperature of 8.0 °C and an annual precipitation of 776.5 mm (1961–1990, German Meteorological Service, station "Landershofen"). Eichstätt is located at the Altmühl river and has a few open green areas. The rectangular-shaped urban park lies south-east of the city center, next to the main campus of the Catholic University of Eichstätt-Ingolstadt (Figure 1c). Hofgarten has an area of 22,480 m² (~2.2 ha). In 1735, the park was designed as a baroque garden and partially transformed to an English garden after 1817. In recent times, these concepts were combined leading to a composition of geometrically designed and accurately pruned baroque elements with long and structured avenues and a variety of some very old trees originating from Europe, North America and Asia [20]. Recently, the park counts 231 trees and shrubs of 69 different species (excluding hedges, flowerbeds and grass and herb species) (Table 1).

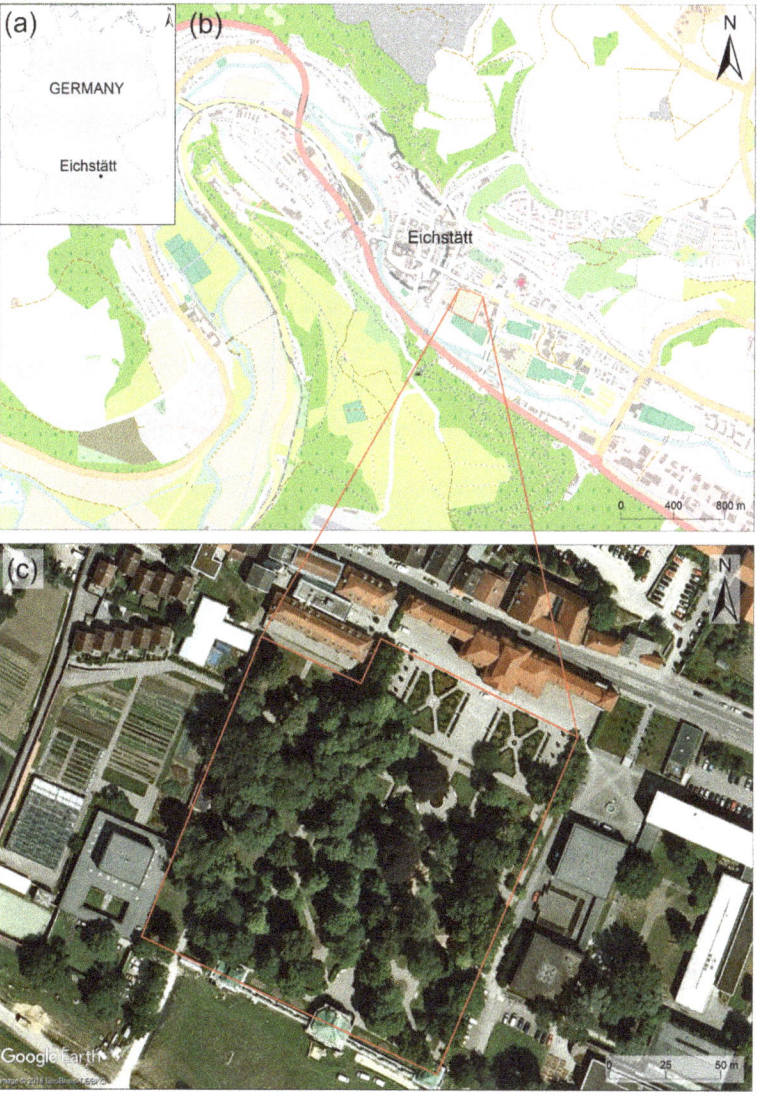

Figure 1. Location of the study site in Germany (Eurostat, NUTS 2013/EU-28) (**a**) and in the city of Eichstätt, red border; (**b**) OpenStreetMap; (**c**) GoogleEarth.

Table 1. Number (N) of individuals planted in Hofgarten belonging to 69 species and 31 plant families (according to the APG system [21]). ap: allergenic potential, pe: type of pollen emission, ppp: duration of principal pollination period (see Table 2 for parameters and values).

Species	N	Family	ap	pe	ppp	Species	N	Family	ap	pe	ppp
Acer griseum (Franch.) Pax 1902	1	Sapindaceae	2	2	1	Liquidambar styraciflua L.	1	Altingiaceae	2	3	1
Acer monspessulanum L.	1	Sapindaceae	2	1	2	Liriodendron tulipifera L.	1	Magnoliaceae	1	1	2
Acer negundo L.	1	Sapindaceae	2	2	1	Magnolia × soulangeana Soul.-Bod.	3	Magnoliaceae	2	1	2
Acer pensylvanicum L. 1753	1	Sapindaceae	2	2	1	Magnolia kobus DC.	1	Magnoliaceae	2	1	2
Acer platanoides L.	19	Sapindaceae	3	2	2	Magnolia stellata (Siebold & Zucc.) Maxim.	3	Magnoliaceae	2	1	2
Acer pseudoplatanus L. 1753	24	Sapindaceae	3	2	1	Morus alba L. 1753	1	Moraceae	2	1	1
Acer rubrum L. 1753	2	Sapindaceae	1	2	2	Nothofagus Antarctica (G. Forster) Oerst.	1	Nothofagaceae	4	3	2
Aesculus × carnea Zeyh.	2	Sapindaceae	2	2	2	Ostrya carpinifolia Scop.	1	Betulaceae	4	3	3
Aesculus hippocastanum L.	18	Sapindaceae	2	2	2	Paeonia × suffruticosa Andrews	1	Paeoniaceae	1	1	2
Ailanthus altissima (Mill.) Swingle	1	Simaroubaceae	3	2	1	Paulownia tomentosa (Thunb.) Steud.	1	Paulowniaceae	2	1	2
Berberis vulgaris L.	1	Berberidaceae	1	1	3	Philadelphus coronarius L.	4	Hydrangeaceae	1	1	3
Betula pendula Roth	2	Betulaceae	4	3	2	Picea omorika (Pančić) Purk.	1	Pinaceae	1	3	1
Buxus sempervirens L.	24	Buxaceae	2	1	2	Picea pungens Engelm.	1	Pinaceae	1	3	3
Carpinus betulus L.	1	Betulaceae	4	3	2	Platanus × hispanica (Aiton) Willd.	1	Platanaceae	3	3	1
Castanea sativa (Mill.)	2	Fagaceae	2	3	1	Potentilla fruticosa (L.) Rydb.	1	Rosaceae	1	1	3
Catalpa bignonioides Walter	1	Bignoniaceae	3	1	2	Prunus padus L.	2	Rosaceae	2	1	2
Celtis australis L.	1	Cannabaceae	3	3	1	Prunus sargentii 'Accolade' Rehder	3	Rosaceae	2	1	2
Cercidiphyllum japonicum Siebold & Zucc.	1	Cercidiphyllaceae	2	1	2	Prunus serrulata 'Kanzan' LINDL.	1	Rosaceae	2	1	2
Cercis siliquastrum L.	1	Fabaceae	2	1	3	Prunus tomentosa Thunb.	1	Rosaceae	1	1	2
Chaenomeles japonica (Thunb.) Lindl. ex Spach	1	Rosaceae	1	1	2	Pterocarya fraxinifolia (Lam.) Spach	1	Juglandaceae	2	3	2
Chamaecyparis lawsoniana (A. Murray) Parl.	1	Cupressaceae	3	3	2	Quercus petraea (Matt.) Liebl.	1	Fagaceae	4	3	1
Chamaecyparis nootkatensis D.Don 1824	1	Cupressaceae	3	3	2	Quercus robur L.	4	Fagaceae	4	3	1
Cornus mas L.	4	Cornaceae	2	1	3	Ribes alpinum L.	2	Grossulariaceae	1	0	2
Corylus avellana L.	1	Betulaceae	4	3	3	Sequoiadendron giganteum (Lindl.) J.Buchh.	2	Cupressaceae	2	3	2
Corylus colurna L.	1	Betulaceae	4	3	2	Sorbus aria (L.) Crantz	1	Rosaceae	1	1	1
Crataegus monogyna Jacq.	1	Rosaceae	1	1	2	Sorbus aucuparia L.	2	Rosaceae	1	1	2
Deutzia scabra Thunb	1	Hydrangeaceae	1	1	2	Sorbus domestica L.	1	Rosaceae	1	1	1
Fagus sylvatica L.	3	Fagaceae	4	3	2	Sorbus torminalis (L.) Crantz	1	Rosaceae	1	1	2
Fraxinus excelsior L.	6	Oleaceae	4	3	2	Spiraea x arguta Zabel	1	Rosaceae	2	1	2
Ginkgo biloba L.	3	Ginkgoaceae	2	3	1	Styphnolobium japonicum (L.) Schott	1	Fabaceae	2	2	2
Gleditsia triacanthos L.	1	Fabaceae	1	0	2	Taxus baccata L.	12	Taxaceae	3	0	3
Hedera helix 'Arborescens' L.	2	Araliaceae	2	1	3	Tilia cordata Mill.	15	Malvaceae	2	2	2
Ilex aquifolium L.	1	Aquifoliaceae	2	0	2	Tilia platyphyllos Scop.	21	Malvaceae	2	2	1
Kolkwitzia amabilis Graebn. Christenh.	1	Caprifoliaceae	1	1	2	Tilia tomentosa Moench	4	Malvaceae	2	2	2
Larix decidua (Mill.)	1	Pinaceae	1	3	3	Sum	231				

2.2. I_{UGZA}—Urban Green Zone Allergenicity Index

The Urban Green Zone Allergenicity Index (I_{UGZA}, [17], Equation (1)) compares an existent green space with a hypothetical space with maximum allergenicity. The index considers a variety of plant characteristics, partially adjusted for our study (Table 2).

$$I_{UGZA} = \frac{1}{378 \cdot S_T} \cdot \sum_{i=1}^{k} n_i \cdot ap_i \cdot pe_i \cdot ppp_i \cdot S_i \cdot H_i \tag{1}$$

Table 2. Scale of values for the parameters used for I_{UGZA} (Urban Green Zone Allergenicity Index) and I_{ISA} (Index of individual-specific allergenic potential of green spaces).

Parameters	Values for I_{UGZA} and I_{ISA}	
Allergenic potential (ap)	0 = non-allergenic (OPALS 1)	
	1 = low allergenicity (OPALS 2–4)	
	2 = moderate allergenicity (OPALS 5–7)	
	3 = high allergenicity (OPALS 8–10)	
	4 = main local allergens	
Type of pollen emissions (pe)	0 = only female-sex individuals	
	1 = entomophilous	
	2 = ampiphilous	
	3 = anemophilous	
Principal pollination period (ppp)	1 = 1–4 weeks	
	2 = 5–8 weeks	
	3 ≥ 9 weeks	
	I_{UGZA}	I_{ISA}
Crown height (H)	Mean height attained at reproductive maturity: 2, 6, 10, 14 m or exceptionally 18 m	Individual-specific measurements [m]
Plant surface (S)	Small-diameter: <4 m, medium-diameter: 4–6 m, large-diameter: >6 m	Individual-specific measurements using 4 radii [m]

OPALS: Ogren Plant Allergy Scale [10].

Relevant variables for I_{UGZA} are the total area of the examined green space in square meters (S_T), the number of species (k), the number of individuals belonging to one species i (n_i), species-specific and classified values for allergenicity (ap_i), type of pollen emission (pe_i), duration of the main pollination period (ppp_i), crown height in meters (H_i) and surface area of the plant in square meters (S_i) (classification see Table 2). The values for the parameters of the index were—apart from the base area of the green space (S_T) and of the present number of species and individuals (k, n_i)—obtained from databases or reports and in this study adapted to conditions prevailing in the biogeographic region of our investigated park (see Section 2.4). The capacity of species-specific pollen emission ($S_i \times H_i$) was calculated using a volume calculation of geometric plant shapes. A cylindrical shape was used for trees and a hemispherical shape for shrubs. The height of the crown (H_i) and surface area of trees and shrubs (S_i) refer to the maximum values of a mature individual of the respective species. For simplification, these literature-based values of H_i and S_i were also classified or scaled.

The figure 378 is an expression of the maximum values ($ap_i \times pe_i \times ppp_i \times H_i$) a species i can obtain and serves, together with the base area of the green space (S_T), as reference unit of the formula. Although, maximum values for height can reach 18 m and main local allergens are considered as ap = 4, [17] used ap = 3, pe = 3, ppp = 3 and H = 14 for the calculation of this factor.

An index value of 0 can only be obtained if the tree is female (pe = 0) or the emitted pollen is non-allergenic (ap = 0).

Values higher than 0.5 already relate to a high allergenic potential [17]. When a densely planted green space is considered, the sum of all surface areas occupied by the plants can be greater than the base area of the green space (S_T). In this case and when all factors and parameters measured are maximal, the index can also exceed the value of 1.

2.3. I_{ISA}—Index of Individual-Specific Allergenic Potential of Green Spaces

The index of the individual-specific allergenic potential of green spaces (I_{ISA}) was calculated using the same formula as for I_{UGZA} (Equation (1)). In contrast to I_{UGZA}, we measured the crown height (H_i) and surface areas (S_i) of each plant. In addition, I_{ISA} uses a different constant (1/1188) since we considered main local allergens (ap = 4) and the maximum tree height differs in our study due to plant-specific measurements. The highest crown represented by the species *Tilia tomentosa* was H = 33 m. In contrast to the study presented by [17] we only included mature individuals that are already producing flowers and hence emitting pollen. For both variants (I_{UGZA} and I_{ISA}), we excluded flowerbeds and herb and grass species and revised all parameters according to the descriptions below.

2.4. Parameters Used for I_{UGZA} and I_{ISA}

For the classification of the allergenic potential (ap_i) of plants, we used the Ogren Plant Allergy Scale (OPALS, [10], reclassification see Table 2). Different species composition and airborne pollen concentrations imply varying sensitisation rates for specific species between countries/regions [8]. Thus, locally occurring highly allergenic plants were additionally taken into account (ap = 4). In our studied park, this relates to species of Betulaceae, Fagaceae and Oleaceae [11]. For missing species not listed in [10], we calculated the median. e.g., the median of all *Acer* species was used to obtain the value for missing information on the allergenic potential of *Acer monspessulanum*.

The factor type of pollen emission (pe_i) consists of information about the type of pollination. E.g., female dioecious plants do not emit any pollen whereas anemophilous plants emit far more pollen than entomophilous plants [22]. The classification of female plants resulted in ap = 0. This classification was mainly made by means of optical characteristics (presence of fruits, e.g., berries on *Ilex aquifolium* = female). If optical inspections were not possible due to plant height, we used the mean of the allergenicity values for female and male plants of the respective species. For the classification of missing species not evaluated by [17], we used information of the databases BiolFlor [23] and Baumkunde.de [24]. In case of diverging specifications, a plant was attributed to pe = 2.

To adopt the length of the pollination period (ppp_i) to the biogeographical region of our study, information on this parameter was obtained from literature and databases [23–26]. Therefore, it was also necessary to adjust the width of classes for this parameter (see Table 2).

The individual-specific volume ($S_i \times H_i$) of each tree is considered to represent the capacity of pollen emission [18]. In contrast to I_{UGZA}, the newly developed I_{ISA} does not incorporate mean values for adult trees but takes the actual information of S_i and H_i for each individual into account. As described by [17], a cylindrical shape was used for trees and a hemispherical shape for shrubs. The unequal symmetry of many trees and shrubs implies that the calculation of the volume of the cylinder remains imprecise when an average radius of the crown is applied. Therefore, the plant's radius was measured at four positions. Tree and crown height was assessed using a height meter (SUUNTO PM-5/1520).

Since I_{ISA} only includes mature individuals, the age of the trees as an indicator for maturity was assessed using a commonly used method proposed by [27] which is based on the measurement of trunk circumferences at a height of 1.5 m. According to [27], the age of the tree is its circumference in centimetres divided by 2.5. In the case of fast-growing trees (e.g., *Sequoiadendron giganteum*, *Pterocarya fraxinifolia*, *Nothofagus antarctica*, *Liriodendron tulipifera*), the circumference is divided by 5. Information regarding species-specific age of maturity was obtained from different publications [28–31]. In case of missing information the mean age of maturity of all available species (=20.1 years) was used as a threshold value of maturity. Note that the growth of solitary trees may differ but could not be assessed in the course of our study.

2.5. Planting Scenarios

Since I_{UGZA} uses the maximum plant height and surface of the crown (values known from literature), but I_{ISA} requires individual-specific values, it was more sensible to use I_{UGZA} in planting scenarios. In all six scenarios, the number of hypothetical planted individuals equals the number of individuals currently growing in the studied park (N = 231).

For scenario 1, we selected 14 typical park trees according to [32,33]. We calculated their count (see numbers in brackets) according to their current proportional occurrence in the park: *Acer plantanoides* (34.3), *Acer pseudoplatanus* (43.3), *Aesculus hippocastanum* (32.5), *Betula pendula* (3.6), *Carpinus betulus* (1.8), *Fagus sylvatica* (5.4), *Fraxinus excelsior* (10.8), *Larix decidua* (1.8), *Picea omorica* (1.8), *Picea pungens* (1.8), *Taxus baccata* (21.7), *Tilia cordata* (27.1), *Tilia platyphyllos* (37.9), *T. tomentosa* (7.2). Scenario 2 is based on the exclusion of locally high-allergenic species. In total 21 trees/shrubs with ap = 4 belonging to the plant families Betulaceae, Fagaceae and Oleaceae were replaced by randomly selected individuals. The median of this new selection was ap = 2. For scenario 3, all present species were selected but uniformly distributed. With 231 individuals and 69 different species, a frequency of 3.35 was assigned to these species. In scenario 4, only the ten most prevalent species were considered (*A. pseudoplatanus*, *T. platyphyllos*, *A. platanoides*, *A. hippocastanum*, *T. cordata*, *T. baccata*, *F. excelsior*, *T. tomentosa*, *Quercus robur*). The number of individuals were uniformly distributed among species. Thus, a frequency of 23.1 was assigned to each species. For scenario 5, also only the ten most prevalent species were incorporated in our calculations, however, with an unequal distribution. The frequencies were 1, 2, 3, 5, 10, 20, 30, 40, 50 and 70 with the lowest frequency related to the species that is represented fewest of all. We considered 24 species in scenario 6 recommended by [34] for city trees and shrubs that were classified at least as suitable in the categories of drought tolerance and winter hardiness. Those climate-tolerant species were uniformly distributed and yielded a count of 9.6. Species belonging to this category were *Acer negundo*, *Sorbus aria*, *Buxus sempervirens*, *Cornus mas*, *Acer rubrum*, *Ailanthus altissima*, *Ginkgo biloba*, *Gleditsia triacanthos*, *Styphnolobium japonicum*, *Sorbus domestica*, *Sorbus torminalis*, *T. tomentosa*, *A. monspessulanum*, *A. platanoides*, *Aesculus x carnea*, *B. pendula*, *C. betulus*, *P. omorika*, *T. cordata*, *Crataegus monogyna*, *Castanea sativa*, *Corylus colurna*, *F. excelsior* and *Quercus petraea*.

2.6. Shannon Index (H_S)

The Shannon index H_S is a commonly used index to describe biodiversity [35] and was used for the comparison of different planting scenarios. If H_S = 0, only one species is prevailing, accounting for 100% of the individuals. The highest diversity represents many different species with an even distribution of individuals. The Shannon index was calculated using Equation (2):

$$H_S = \sum_{i=1}^{S} p_i \cdot \ln(p_i) \text{ with } p_i = \frac{n_i}{N} \tag{2}$$

where S = number of species, p_i = relative frequency of the ith species, n_i = number of individuals belonging to one species, and N = number of individuals.

3. Results

3.1. Plant Characteristics and Current Allergenic Potential of the Park

The allergenic potential of the investigated 231 trees is predominantly moderate (53%; Figure 2a). Only a small number of trees (11%) was characterised as low allergenic and a comparable amount (9%) is even considered as a source of main local allergens (e.g., *B. pendula* or *Corylus avellana*). The most common pollination strategy (48%) is the mixture of entomophily and anemophily (Figure 2b). Only 7% of the individuals were female plants producing no pollen (e.g., *T. baccata* or *I. aquifolium*). The majority of trees (59%) have a principal pollination period of 4–6 weeks; merely 12% of the plant species flower for 9 weeks or even longer (Figure 2c).

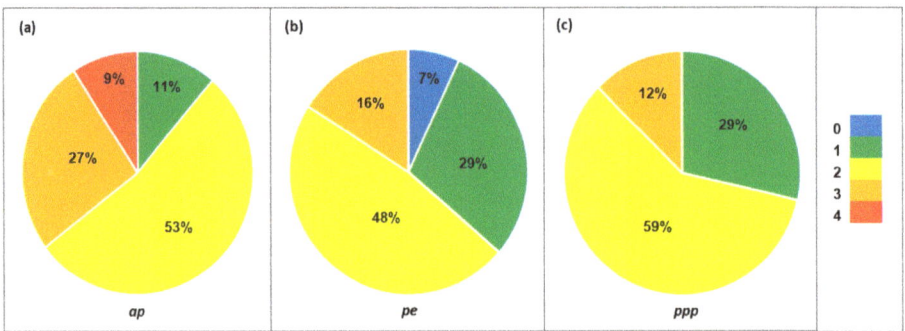

Figure 2. Percentage of plant individuals (N = 231) in the urban park Hofgarten classified according to their (**a**) allergenic potential (ap); (**b**) type of pollen emission (pe) and (**c**) duration of principal pollination period (ppp, see Table 2 for parameters and values).

The allergenic potential of the park calculated using literature values for crown height and surface was I_{UGZA} = 0.173. Assuming no alterations of planted species, the park's potential is limited to a low allergenic potential. Applying our own measurements to the data reduced the index to I_{ISA} = 0.018. Thus, the current allergenic potential is substantially overestimated using I_{UGZA}.

3.2. Planting Scenarios and Associated Biodiversity

Although some species producing pollen known as main local allergens (e.g., pollen of *B. pendula* and *F. excelsior*) were considered in scenario 1, the planting of typical park trees [32,33] according to their current proportional occurrence was associated with the highest allergenic potential (I_{UGZA} = 0.226, see Table 3). Replacing trees considered as sources of main local allergens (scenario 2) had the strongest positive impact on the park's allergenic potential (I_{UGZA} = 0.147), almost comparable to scenario 5 (I_{UGZA} = 0.150) where the ten most predominant species with an unequal distribution were selected. An equal distribution, however, induced a comparable high value (I_{UGZA} = 0.197, scenario 4). Applying scenario 3 (uniform distribution of all present species) or scenario 6 (climate-tolerant species [34]), yielded quite similar values for I_{UGZA} compared to the current state (see Table 3).

A greater biodiversity (according to Shannon index H_s) was not necessarily linked to a lower allergenic potential of the park (Table 3 and Figure 3). In fact, the highest biodiversity (H_s = 4.23) in scenario 3 (uniform distribution of all present species) only yielded an average allergenic potential (I_{UGZA} = 0.170). The lowest biodiversity (H_s = 1.81) in scenario 5 (unequal distribution of ten most occurring species) was even associated with a relatively low allergenic potential (I_{UGZA} = 0.150).

Table 3. Urban Green Zone Allergenicity Index (I_{UGZA}) for the current state of the urban park (0) and for current state and different planting scenarios and their corresponding Shannon index (H_s).

	I_{UGZA}	H_s
Current state (0)	0.173	3.47
Scenario 1	0.226	2.20
Scenario 2	0.147	3.39
Scenario 3	0.170	4.23
Scenario 4	0.197	2.30
Scenario 5	0.150	1.81
Scenario 6	0.171	3.18

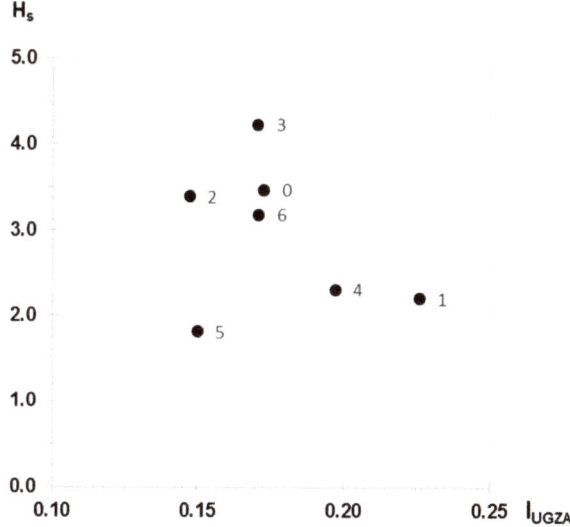

Figure 3. Scatterplot with the variables Shannon index (H_s) and Urban Green Zone Allergenicity Index (I_{UGZA}) for the current state of the urban park (0) and for six planting scenarios (values see Table 3).

4. Discussion

4.1. Comparison of Indices and Conceptual Remarks

The indices I_{UGZA} and I_{ISA} used in our study can be used as a management tool in urban green planning. These indices offer a comparison of the allergenic potential of different parks or planting scenarios, therefore, they can contribute to the development of allergy-friendly parks. Since the presented indices account for the total area of the examined green space (see Equation (1)), the methods presented in our study are applicable for any park of any size. In fact, recommendations for planting trees can be applied with little effort, leading to positive effects for pollen allergy sufferers.

The I_{UGZA} facilitates an easy way to assess and predict the allergenic potential of an urban green space by the use of only a few mathematical calculations. To assess the individual-specific allergenic potential of urban green spaces, we developed an index (I_{ISA}) that includes individual-specific foliage volume by accurate measurements of crown heights and surface areas occupied by each plant. Since there exists an apparent link between foliage volume and number of flowers (and therefore intensity of pollen emission [18]), it is sensible to use individual-specific measurements. Therefore, the I_{ISA} has advantages such as the improvement in accuracy and the reduction of a high influence of the parameters height and surface area. On the other hand, there are constraints in practicability and in speed of the analysis.

In general, the I_{ISA} resulted in a lower value compared to I_{UGZA} (0.018 vs. 0.173). This difference is not surprising because the data obtained for calculating I_{UGZA} are based on mean values of mature individuals (which is different to the individual-specific measurement). The actual growth rate of trees is dependent on different site-specific factors such as climate, competition or nutrient supply [36]. Therefore, the calculation of I_{UGZA} based on literature data may lead to an overestimation of the actual crown volume and in turn to an underestimation of the intrinsic allergenic potential. This fact was also mentioned by [37] who applied this index to three localities in Spain.

The purpose of both indices is to compare an actual green space with a hypothetical green space with maximum allergenicity. Generally, plant crowns might overlap resulting in a higher surface area occupied by plants compared to the geometrical surface area of the park. Thus, a conceptual weakness of both indices is the fact that values can also be higher than 1 (see Methods section). In reality, however, the indices are rather small for parks (see Results [17,37]), but not for dense forests [38]. Even smaller values would be obtained when the constant of the formula for I_{UGZA} is adjusted to 648 (by applying the prevailing maximum values for the parameters used: ap = 4, pe = 3, ppp = 3 and H = 18). The value for I_{ISA} (0.018) was considerably smaller since the constant used in the equation (1188) included sources of main local allergens (ap = 4) and was adapted to the maximum measured crown height (H = 33 m, *T. tomentosa*). Some species, e.g., *C. avellana*, will never reach a comparable height. Therefore, this index might yield in more meaningful values when the height of trees does not differ very much among planted trees or shrubs.

Another modification was the reduction of the number of considered individuals due to the fact that only mature species were included for the calculation of I_{ISA}. Considering all species (both mature and immature), the index was quite similar (0.019, data not shown). This is probably attributable to the fact that immature individuals are smaller in size and therefore do not contribute much to the magnitude of the index.

Allergenic pollen is not only restricted to trees and shrubs. The high sensitisation rate in German adults of 37.9% linked to grass pollen [39] shows the importance of lawns. In our studied park, grass is cut on a regular basis and only some small areas covering less than 5% of the park's area are not cut to promote spontaneous vegetation for insects. Thus, this study does not include the allergenic potential of non-woody plants but only the allergenic potential based on trees and shrubs. The inclusion of lawn area which accounts for 55.9% of the park area (1.3 ha) would result in an increase of I_{UGZA} by 0.00254 when a medium height of 0.2 m is assumed. Including grass for I_{ISA} is not sensible since this index is based on accurate measurements of plant heights (which varies in the course of the vegetation season in the case of grass) and only includes mature individuals. Most grasses, however, do not flower in our park due to frequent cutting. Nevertheless, in other urban green areas, the inclusion of non-woody plants may has to be considered as well.

The employment of pruning can be an effective management tool. In our park, accurately pruned individuals of *B. sempervirens* expose a lower risk since flowers are kept to a minimum. Ideally, information on pruning practices should be included in assessing the allergenic potential of parks. The studies of [17,37] accounted for the effects of grass and flower species. Here, we excluded hedges and lawns since they are very frequently cut.

An important factor not considered in our study is the pollen emission of areas outside the study area [8]. In case of Hofgarten, an inflow especially from south-eastern (plantings and lawns), south-western (large green areas and riverside vegetation) and north-western directions (plant nursery) is possible, despite walls and buildings enclosing this park. Depending on species and wind conditions, pollen can be spread over large areas and great distances [40,41] substantially influencing the allergenic potential of a given place. Pollen abundance is further affected by numerous surfaces for impaction and filtration [42] such as walls and buildings that surround Hofgarten. In addition, needles of conifers or leaves of deciduous trees may influence the impaction of pollen. Early flowering anemophilous plants such as hazel and birch may be able to disperse their pollen more efficiently since their own leaves and most leaves of other plants are not unfolded yet. Except for some fountains, lakes or ponds are not present in Hofgarten, but they allow the deposition of pollen [38]. Since the length of the pollination period (ppp) may vary due to different weather conditions from one year to another [43], phenological observations and/or airborne pollen measurement might contribute to a more accurate index in further studies. In the study of [37], pollen concentrations were monitored at the roof of high buildings, but not in the investigated parks. An exact assessment of the influence of pollen transport however, is only possible with on-site airborne pollen measurements.

4.2. Planting Scenarios and Recommendations for Plantings in Urban Green Areas

The selection of 14 typical park trees in German cities [32,33] according to their current proportional occurrence resulted in the highest allergenic potential (I_{UGZA} = 0.226) among all six planting scenarios. This finding suggests that the planting of common park trees in general is not very suitable for designing allergy-friendly green spaces. This may be attributed to some species whose pollen are known as main local allergens such as *B. pendula* or *F. excelsior*. Regarding the current state of the park, a comparison between these typical park trees (93 individuals) and the remaining trees and shrubs which are not very common or even exotic (138 individuals) shows that their proportional I_{UGZA} is almost equal (0.087 vs. 0.086; data not shown). This finding also suggests that typical park trees include species with negative effects for allergy sufferers. It was pointed out that a moderate planting of exotic species leads to an increasing floral diversification, but the overuse of these exotics should be avoided [14]. *Olea europaea* is an example of a species producing highly allergenic pollen that is mainly planted because of its exotic appearance [10]. Frequently planted individuals of *O. europaea* lead to an increase of airborne Oleaceae pollen and therefore to an increase of the sensitisation rate of German adults which is currently at 9.7% [39]. This species might pose a risk since a high degree of family relationship with another more common species with allergenic pollen (*Fraxinus* spp.) exists [44]. The role of exotic species is controversially debated: Whereas [45] attribute exotic species a preventive measure to reduce sensitisations, other authors noted that some species attracted negative attention. For southern Spain, [46] mentioned allergy symptoms in autumn related to the genus *Casuarina* that is native to Australia and Asia but extensively used as an ornamental tree, especially in coastal cities [47].

The role of main local allergens for the allergenic potential of green spaces is evident: Regarding the current state of the park, 21 major local allergenic trees and shrubs are contributing with a value of 0.036 to the allergenic potential (data not shown). This means that only 9% of the trees and shrubs are responsible for 21.1% of the allergenic potential calculated with I_{UGZA}. For example, major local allergenic pollen of birch, hazel and grasses are triggering clinically relevant symptoms in over 90% of the German adults and on over 75% of Europeans [39]. Thus, replacing main local allergenic plants (scenario 2) had a considerable impact on the park's allergenic potential (I_{UGZA} = 0.147). The substitution was made by randomly selected individuals with a medium allergenic potential of ap = 2. Using solely non-allergenic plants would yield in even lower values of I_{UGZA}.

To assess the suitability of the current selection of trees, we incorporated three different scenarios. Although a uniform distribution of all planted species (scenario 3) was linked to a similar I_{UGZA} as the actual selection, an equal distribution of the ten most frequently occurring plants (scenario 4) resulted in a comparably high allergenic potential (I_{UGZA} = 0.197). This indicates that the actual preferential selection of trees is not very suitable with respect to their allergenic potential. An unequal distribution (scenario 5, I_{UGZA} = 0.156) reduced the high influence on I_{UGZA} of some species such as the main local allergenic trees *Q. robur* and *F. excelsior* as well as the high allergenic potential of *T. tomentosa* because of its great foliage volume. As a result, the mass use of certain plants results in large quantities of monospecific pollen, probably affecting the frequency of new sensitisations and the aggravation of symptoms in people allergic to this pollen [14].

The categorisation of climate-tolerant species by [34] does not include any parameters related to the risk for allergy sufferers. The species that are able to perform well under changed climatic conditions (scenario 6) yielded similar values for I_{UGZA} compared to the current state. However, for these and for all other species, it has to be considered that plant-related factors such as the length of the pollination period or the pollen's allergenicity might change with ongoing climate change [48]. In the future, these changes might result in different assessments and have to be timely regarded in urban planning.

The results of our planting scenarios suggest that a greater diversity of trees, assessed using Shannon index, is not linked to a lower allergenic potential of the park. However, several studies claim that urban green planning should focus on biodiversity for ameliorating the allergenic potential of parks [14,17]. Additional species are increasing species diversity but we suppose that the mass use of common pollen emitters leads to high pollen levels increasing the risk of new sensitisations and aggravating allergic reactions.

Allergy sufferers should have access to green spaces where they are not put at additional health risk. In Germany, pollen allergens can be found nearly year-round in the air [26], which can strongly affect those concerned. The occurrence of highly allergenic and anemophilous plants should be limited or avoided, because of their high pollen emission [8]. Furthermore, plants with a long pollination period and the increased planting of only male individuals of dioecious species pose a risk for allergy sufferers. It was suggested that the concept of a female park is suitable for allergy sufferers, because female individuals of dioecious species do not release any pollen [10]. However, the sterility of such green spaces results in a minimised capability of pollination [8]. In addition, female trees are often not favoured for planting due to higher amounts of litter or undesirable odour as observed in *G. biloba* [14].

5. Conclusions

In order to keep and design cities as liveable as possible, it is important to regard urban green spaces as providers for not only positive ecosystem services but possibly also for *ecosystem disservices* related to allergies. Until now, not many specific recommendations for the implementation of allergy-friendly city greening exist and are therefore needed. Indices are a useful management tool for analysing and assessing allergenic potential in public green spaces. With purposive consideration of single individuals, they allow to make an actual site-specific assessment. Based on literature values and a few mathematical equations the I_{UGZA} is easy to calculate. The individual-specific allergenic potential of urban green spaces assessed by I_{ISA} requires more information obtained by accurate measurements of crown heights and surface areas occupied by each plant. Thus, the calculation of this index is more time consuming but the result is more precise. Planting scenarios and assessing biodiversityare useful for the formulation of recommendations for urban planning. In further studies, influences such as pollen flow from neighbouring or distant green areas should be taken into account.

Author Contributions: S.J.-O. and T.S. conceived and designed the study; T.S. was involved in field measurements; J.J. prepared maps of the study area; S.J.-O. analysed the data; P.C. contributed with methods and plant information; S.J.-O. wrote the paper with the help of all co-authors.

Acknowledgments: We gratefully acknowledge the support of the Spanish Ministry of Economy and Competitiveness (FENOMED CGL2014-54731-R).

Conflicts of Interest: The authors declare no conflict of interest.

References

1. Bergmann, K.-C.; Heinrich, J.; Niemann, H. Current status of allergy prevalence in Germany: Position paper of the Environmental Medicine Commission of the Robert Koch Institute. *Allergo J. Int.* **2016**, *25*, 6–10. [CrossRef] [PubMed]
2. Ring, J.; Bachert, C.; Bauer, C.-P.; Czech, W. *Weißbuch Allergie in Deutschland*, 3rd ed.; Springer Medizin: München, Germany, 2010, ISBN 978-3-89935-245-0. (In German)
3. Langen, U.; Schmitz, R.; Steppuhn, H. Häufigkeit allergischer Erkrankungen in Deutschland: Ergebnisse der Studie zur Gesundheit Erwachsener in Deutschland (DEGS1). *Bundesgesundh. Gesundheitsforsch. Gesundheitssch.* **2013**, *56*, 698–706. (In German) [CrossRef] [PubMed]
4. Bundesministerium für Umwelt, Naturschutz, Bau und Reaktorsicherheit. *Grün in der Stadt—Für eine Lebenswerte Zukunft. Grünbuch Stadtgrün*; BMUB: Berlin, Germany, 2015. (In German)
5. Millennium Ecosystem Assessment. *Ecosystems and Human Well-Being: Synthesis*; Island Press: Washington, DC, USA, 2005, ISBN 1-59726-040-1.
6. Lyytimäki, J.; Sipilä, M. Hopping on one leg—The challenge of ecosystem disservices for urban green management. *Urban For. Urban Green.* **2009**, *8*, 309–315. [CrossRef]
7. Baumschulen Gebr. van den Berk. *Van den Berk über Bäume*, 2nd ed.; Baumschule Van den Berk: Sint-Oedenrode, Germany, 2004, ISBN 90-807408-6-1. (In German)
8. Seyfang, V. *Studie zum Wissenschaftlichen Erkenntnisstand über das Allergiepotential von Pollenflug der Gehölze im Öffentlichen Grün der Städte und Gemeinden und Mögliche Minderungsstrategien*; Hochschule Ostwestfalen-Lippe: Lemgo, Germany, 2008. (In German)

9. Huntington, L. *Das Gartenbuch für Allergiker. Die Schönsten Pflanzen, die Besten Arbeitsweisen*; VGS: Köln, Germany, 1999, ISBN 3802513851. (In German)
10. Ogren, T.L. *Allergy-Free Gardening. The Revolutionary Guide to Healthy Landscaping*; Ten Speed Press: Berkeley, CA, USA, 2000, ISBN 1580081665.
11. Bergmann, K.-C.; Zuberbier, T.; Ausgustin, J.; Mücke, H.-G.; Straff, W. Klimawandel und Pollenallergie. Städte und Kommunen sollten bei der Bepflanzung des öffentlichen Raums Rücksicht auf Pollenallergiker nehmen. *Allergo J.* **2012**, *21*, 103–108. (In German) [CrossRef]
12. Sandifer, P.A.; Sutton-Grier, A.E.; Ward, B.P. Exploring connections among nature, biodiversity, ecosystem services, and human health and well-being: Opportunities to enhance health and biodiversity conservation. *Ecosyst. Serv.* **2015**, *12*, 1–15. [CrossRef]
13. Hanski, I.; von Hertzen, L.; Fyhrquist, N.; Koskinen, K.; Torppa, K.; Laatikainen, T.; Karisola, P.; Auvinen, P.; Paulin, L.; Mäkelä, M.J.; et al. Environmental biodiversity, human microbiota, and allergy are interrelated. *Proc. Natl. Acad. Sci. USA* **2012**, *109*, 8334–8339. [CrossRef] [PubMed]
14. Cariñanos, P.; Casares-Porcel, M. Urban green zones and related pollen allergy: A review. Some guidelines for designing spaces with low allergy impact. *Landsc. Urban Plan.* **2011**, *101*, 205–214. [CrossRef]
15. Hruska, K. Assessment of urban allergophytes using an allergen index. *Aerobiologia* **2003**, *19*, 107–111. [CrossRef]
16. Ciferri, E.; Torrisi, M.; Staffolani, L.; Hruska, K. Ecological study of the urban allergenic flora of central Italy. *J. Mediter. Ecol.* **2006**, *7*, 15–21.
17. Cariñanos, P.; Casares-Porcel, M.; Quesada-Rubio, J.-M. Estimating the allergenic potential of urban green spaces: A case-study in Granada, Spain. *Landsc. Urban Plan.* **2014**, *123*, 134–144. [CrossRef]
18. Friedman, J. The Ecology and Evolution of Wind Pollination. Ph.D. Thesis, University of Toronto, Toronto, ON, Canada, 2009.
19. Givnish, T.J. Ecological constraints on the evolution of breeding systems in seed plants: Dioecy and dispersal in Gymnosperms. *Evolution* **1980**, *34*, 959. [CrossRef] [PubMed]
20. Besl, M. Der Hofgarten vor der Sommerresidenz Eichstätt. Genese und Vergleiche. Master's Thesis, Catholic University of Eichstätt-Ingolstadt, Eichstätt, Germany, 2011. (In German)
21. Byng, J.W.; Chase, M.W.; Christenhusz, M.J.; Fay, M.F.; Judd, W.S.; Mabberley, D.J.; Sennikov, A.N.; Soltis, D.E.; Soltis, P.S.; Stevens, P.F.; et al. An update of the Angiosperm Phylogeny Group classification for the orders and families of flowering plants: APG IV. *Bot. J. Linn. Soc.* **2016**, *181*, 1–20. [CrossRef]
22. Pütz, N. *Studienhilfe Botanik*; Vechtaer fachdidaktische Forschungen und Berichte No. 17; Institut für Didaktik der Naturwissenschaften, der Mathematik und des Sachunterrichts: Vechta, Germany, 2008. (In German)
23. Umweltforschungszentrum Leipzig Halle. BiolFlor: Datenbank Biologisch-Ökologischer Merkmale der Flora von Deutschland. Available online: http://www2.ufz.de/biolflor/index.jsp (accessed on 20 February 2017). (In German)
24. Gurk, C.; Hepp, C. Baumkunde.de: Online-Datenbank für Bäume und Sträucher. Available online: http://www.baumkunde.de/ (accessed on 20 February 2017). (In German)
25. Jäger, E.J.; Müller, G.K.; Ritz, C.; Welk, E.; Wesche, K. *Rothmaler—Exkursionsflora von Deutschland. Gefäßpflanzen: Atlasband*, 11th ed.; Springer Spektrum: München, Germany, 2007. (In German)
26. Stiftung Deutscher Polleninformationsdienst. Pollenflugkalender für Deutschland: Pollenflug Gesamtdeutscher Raum 2007–2011. Available online: http://www.pollenstiftung.de/pollenvorhersage/pollenflug-kalender/ (accessed on 7 March 2017). (In German)
27. Mitchell, A. *Die Wald- und Parkbäume Europas. Ein Bestimmungsbuch für Dendrologen und Naturfreunde*, 2nd ed.; Parey: Hamburg, Germany, 1979, ISBN 3490059182. (In German)
28. Bartels, H. *Gehölzkunde*; Ulmer: Stuttgart, Germany, 1993. (In German)
29. Schütt, P.; Weisgerber, H.; Schuck, H.J.; Lang, U.M.; Stimm, B.; Roloff, A. *Enzyklopädie der Laubbäume. Die Große Enzyklopädie mit über 800 Farbfotos unter Mitwirkung von 30 Experten*; Nikol: Hamburg, Germany, 2006, ISBN 3937872396. (In German)
30. Schütt, P.; Schuck, H.J.; Stimm, B. *Lexikon der Bauch- und Strauchbaumarten. Das Standardwerk der Forstbotanik*, 3rd ed.; Nikol: Hamburg, Germany, 2014. (In German)
31. Stinglwagner, G.K.F.; Haseder, I.E.; Erlbeck, R. *Das Kosmos Wald- und Forstlexikon*, 2nd ed.; Kosmos: Stuttgart, Germany, 2005, ISBN 3440103757. (In German)

32. Kunick, W. *Biotopkartierung—Landschaftsökologische Grundlagen*; Teil 3; Stadt Köln: Köln, Germany, 1983. (In German)
33. Wittig, R. *Siedlungsvegetation. 40 Tabellen*; Ulmer: Stuttgart (Hohenheim), Germany, 2002, ISBN 9783800136933. (In German)
34. Roloff, A.; Bonn, S.; Gillner, S. Konsequenzen des Klimawandels—Vorstellung der Klima-Arten-Matrix (KLAM) zur Auswahl geeigneter Baumarten. *Stadt+Grün* **2009**, *57*, 53–60.
35. Munk, K. *Ökologie—Evolution*; Thieme: Stuttgart, Germany, 2009, ISBN 3131448814. (In German)
36. Speer, J.H. *Fundamentals of Tree-Ring Research*; University of Arizona Press: Tuscon, AZ, USA, 2010.
37. Maya Manzano, J.M.; Tormo Molina, R.; Fernández Rodríguez, S.; Silva Palacios, I.; Gonzalo Garijo, Á. Distribution of ornamental urban trees and their influence on airborne pollen in the SW of Iberian Peninsula. *Landsc. Urban Plan.* **2017**, *157*, 434–446. [CrossRef]
38. Cariñanos, P.; Adinolfi, C.; de Guardia, C.L.D.; de Linares, C.; Casares-Porcel, M. Characterization of allergen emission sources in urban areas. *J. Environ. Qual.* **2016**, *45*, 244–252. [CrossRef] [PubMed]
39. Burbach, G.J.; Heinzerling, L.M.; Edenharter, G.; Bachert, C.; Bindslev-Jensen, C.; Bonini, S.; Bousquet, J.; Bousquet-Rouanet, L.; Bousquet, P.J.; Bresciani, M.; et al. GA(2)LEN skin test study II: Clinical relevance of inhalant allergen sensitizations in Europe. *Allergy* **2009**, *64*, 1507–1515. [CrossRef] [PubMed]
40. Stanley, R.G.; Linskens, H.F. *Pollen. Biologie, Biochemie, Gewinnung und Verwendung, Lizenzausg*; Freund: Greifenberg, Germany, 1985, ISBN 9783924733001.
41. Makra, L.; Matyasovszky, I.; Tusnády, G.; Wang, Y.; Csépe, Z.; Bozóki, Z.; Nyúl, L.G.; Erostyák, J.; Bodnár, K.; Sümeghy, Z.; et al. Biogeographical estimates of allergenic pollen transport over regional scales: Common ragweed and Szeged, Hungary as a test case. *Agric. For. Meteorol.* **2016**, *221*, 94–110. [CrossRef]
42. Emberlin, J.; Norris-Hill, J. Spatial variation of pollen deposition in North London. *Grana* **1991**, *30*, 190–195. [CrossRef]
43. Jochner, S.C.; Beck, I.; Behrendt, H.; Traidl-Hoffmann, C.; Menzel, A. Effects of extreme spring temperatures on urban phenology and pollen production: A case study in Munich and Ingolstadt. *Clim. Res.* **2011**, *49*, 101–112. [CrossRef]
44. Vara, A.; Fernández-González, M.; Aira, M.J.; Rodríguez-Rajo, F.J. Fraxinus pollen and allergen concentrations in Ourense (South-western Europe). *Environ. Res.* **2016**, *147*, 241–248. [CrossRef] [PubMed]
45. Chiesura, A. The role of urban parks for the sustainable city. *Landsc. Urban Plan.* **2004**, *68*, 129–138. [CrossRef]
46. García, M.; Moneo, I.; Audicana, M.T.; del Pozo, M.D.; Muñoz, D.; Fernández, E.; Díez, J.; Etxenagusia, M.A.; Ansotegui, I.J.; Fernández de Corres, L. The use of IgE immunoblotting as a diagnostic tool in Anisakis simplex allergy. *J. Allergy Clin. Immunol.* **1997**, *99*, 497–501. [CrossRef]
47. Trigo, I.F.; Davies, T.D.; Bigg, G.R. Objective climatology of cyclones in the Mediterranean region. *J. Clim.* **1999**, *12*, 1685–1696. [CrossRef]
48. Beggs, P.J. Impacts of climate change on aeroallergens: Past and future. *Clin. Exp. Allergy* **2004**, *34*, 1507–1513. [CrossRef] [PubMed]

© 2018 by the authors. Licensee MDPI, Basel, Switzerland. This article is an open access article distributed under the terms and conditions of the Creative Commons Attribution (CC BY) license (http://creativecommons.org/licenses/by/4.0/).

Communication

Tree Vitality Assessment in Urban Landscapes

David Callow [1], Peter May [2] and Denise M. Johnstone [3],*

[1] Urban Forest & Ecology, City of Melbourne, Melbourne VIC 3000, Australia; david.callow@melbourne.vic.gov.au
[2] School of Ecosystem and Forest Sciences, Burnley Campus, University of Melbourne, Richmond VIC 3121, Australia; pmay@unimelb.edu.au
[3] School of Ecosystem and Forest Sciences, Parkville Campus, University of Melbourne, Parkville VIC 3010, Australia; denisej@unimelb.edu.au
* Correspondence: denisej@unimelb.edu.au; Tel.: +61-427-689-495

Received: 28 March 2018; Accepted: 16 May 2018; Published: 21 May 2018

Abstract: The recent prolonged drought in Melbourne, Australia has had a deleterious effect on the urban forest, resulting in the premature decline of many mature trees and a consequent decline in the environmental services that trees are able to provide to urban residents. Measuring the severity of tree stress and defoliation due to various climatic factors is essential to the ongoing delivery of environmental services such as shade and carbon sequestration. This study evaluates two methods to assess the vitality of drought stressed Elm trees within an inner-city environment—bark chlorophyll fluorescence measured on large branches and an urban visual vitality index. Study species were *Ulmus procera* Salisb. (English Elm) and *Ulmus* × *hollandica* (Dutch Elm), which are important character and shade tree species for Melbourne. Relationships were identified between leaf water potential and the urban visual vitality index and between leaf water potential and bark chlorophyll fluorescence measured on large branches, indicating that these methods could be used to assess the effect of long-term drought and other stressors on urban trees.

Keywords: urban tree growth; climate change; drought stress

1. Introduction

The recent prolonged drought in Melbourne, Australia (1997–2010) has had a deleterious effect on many trees within the urban forest. Within the City of Melbourne (a municipality of 37.7 km² around the Melbourne CBD), prolonged drought and resulting mandatory changes to irrigation practices have led to a premature loss of vitality in many mature trees. Climate-related tree decline within the City of Melbourne was particularly evident in exotic deciduous trees, such as *Ulmus* spp., which are widely planted throughout the City of Melbourne and are recognized as an important element of the city landscape [1].

Tree vitality can be defined in relation to the plant response to physiological stress [2]. Very low vitality trees will not respond to treatment to ameliorate physiological stress, but will remain in a depleted state or die, possibly because of extremely low carbohydrate reserves. High vitality trees, on the other hand, will respond and recover from drought or other physiological stressors.

Chlorophyll fluorescence measurements and in particular ratios of F_v/F_m are used to examine aspects of plant photosynthetic and photochemical processes that give rise to plant vitality [3]. However, most chlorophyll fluorescence and many other physiological measurements of plant stress utilize leaf material, which precludes the year-round assessment of deciduous trees and can confound tree vitality with the health of individual leaves [4]. Using chlorophyll fluorescence measurements in bark tissue to assess vitality, though possible, has seldom been used in practice [4]. The most commonly used chlorophyll fluorescence measurement is F_v/F_m, where F_v is the difference between maximum

(F_m) and minimum (F_0) fluorescence [5]. F_v/F_m is the theoretical measure of the quantum efficiency of photosystem two (PSII) during photosynthesis if all the PSII reaction centers are open. A decrease in the chlorophyll fluorescence parameter F_v/F_m indicates photoinhibitory damage. F_v/F_m values can decrease significantly with salt, heat, and herbicide damage [3], as well as with drought stress [6]. Leaf F_v/F_m values between 0.78 to 0.85 are typical for healthy, non-stressed trees [3]; however, bark values are typically lower, probably due to a decreased efficiency in bark chlorenchyms and an external bark layer that may inhibit the fluorescence signal passing to the measurement instrument [4].

Visual methods for the assessment of tree vitality have been successfully applied in natural forest stands [7,8], hardwood plantations [4], and urban environments [9,10]. A numerical crown assessment technique was developed for living *Eucalyptus* (*Corymbia*) *maculata* (Hook.) K.D. Hill & L.A.S. Johnson (spotted gum), *Eucalyptus fibrosa* F. Muell. (ironbark) and *E. drepanophylla* F. Muell ex Benth (syn. *E. cerebra* F. Muell) trees by Grimes [7]. The method used crown position in relation to other trees, crown size, crown density, the number of dead branches, and epicormic growth and developed a prediction equation for how well these factors explained incremental growth at breast height. He found that each of the five variables contributed significantly to a prediction equation for incremental growth, but that for the best results, factors should be weighted differently, for example, epicormic growth on a three-point scale and crown density on a nine-point scale. Martin et al. [8] used the above method by Grimes [7] to develop a visual assessment method to examine the spatial representation of eucalypt dieback across two sites in a natural forest stand. At one site, they identified a positive relationship between vitality and tree size. Aiming to assess the vitality of drought stressed and declining mature trees, Johnstone et al. [4] adapted the method described by Martin et al. [8] for a study in plantation eucalypts and found correlations between visual vitality and total leaf area, above ground biomass, leaf chlorophyll fluorescence (summer), and trunk bark chlorophyll fluorescence (spring, summer, and autumn).

A more detailed discussion of methods to assess long-term drought and/or vitality in trees can be found in Johnstone et al. [2]. In this paper, the authors discuss the relative merits of measuring tree vitality using tree growth and visual parameters; leaf or needle morphology and biochemistry; canopy transparency and reflectance; electrical admittance/impedance; gaseous exchange; and chlorophyll fluorescence.

We report here on how we have used an urban visual vitality index and bark chlorophyll fluorescence measured on large branches to predict drought stress in mature urban trees. Specifically, we tested if there was a relationship between the drought stress in *Ulmus procera* Salisb. (English Elm) and *Ulmus* × *hollandica* (Dutch Elm) and urban visual vitality index values and/or bark chlorophyll fluorescence measurements on large branches. Our results inform management decisions for trees in urban environments for the benefit of city dwellers and users by maximizing canopy cover for the successful provision of environmental services. The results also help tree managers provide measures for the successful establishment of succession planning for tree replacement.

2. Materials and Methods

For this study, trees were examined around the central business district of the City of Melbourne, Australia (Latitude 37°47′ S, longitude 144°58′ E). Melbourne has had an average annual rainfall of 648 mm and high summer temperatures with a January mean maximum of 26 degrees Celsius from 1855 to 2015 [11]. The experimental trees consisted of six mature *Ulmus procera* located in a suburban streetscape in a contiguous row and 32 randomly selected *Ulmus* × *hollandica* located within inner city parks from two avenues. All trees displayed visual drought stress symptoms and displayed variation in crown condition from very stressed to not particularly stressed, to properly evaluate the relationship between drought stress and the two assessment methods.

The urban visual vitality index used (Figure 1) was adapted from a method described by Johnstone et al. [4]. The urban visual vitality index had three criteria: crown size, crown density, and crown epicormic growth. Each of the criteria was scored numerically; 1 to 5 (crown size), 1–9 (crown density), and 1–3 (crown epicormic growth). The relative weighting of each parameter follows Grimes [6] who, as previously stated, developed a predictive equation for how well each factor explained incremental growth at breast height. The score was totaled, with a higher score indicating greater vitality (Figure 1). Urban visual vitality index assessments were undertaken on all trees in December 2010, January 2011, and February 2011. The lowest score in December was 3.5 and the highest was 14.5 out of a possible 17, confirming the range of tree crown conditions mentioned above.

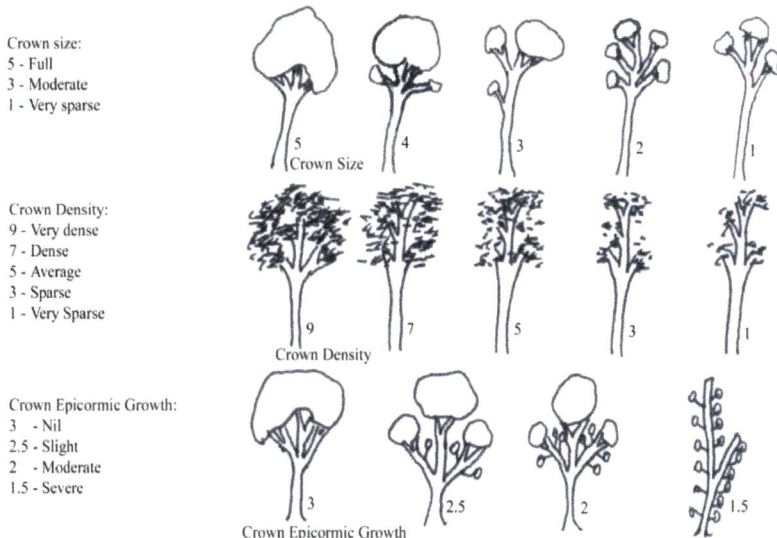

Figure 1. Diagrammatic representation of the assessment of urban visual vitality index for mature trees (after [4,7,8,12]).

Leaf water potential was measured in March 2011 by harvesting five fully expanded sun leaves from within the upper half of the crown of each tree. Leaves were collected using an Elevating Work Platform. For pre-dawn measurements, leaves were collected between 5:30 and 7:00 a.m., prior to direct sunlight reaching any part of the canopy on consecutive days with similar weather conditions. For midday measurements, leaves were collected between 12:00 p.m. and 1:30 p.m. also on consecutive days with similar weather conditions. Water potential readings were undertaken in a PMS Model 1000 pressure chamber (PMS instrument company, Albany, OR, USA) within 20 min of leaf harvesting. Leaf water potential was measured using the protocol first described by Scholander [13].

Bark chlorophyll fluorescence was also measured in March 2011 with a Hansatech-Handy Plant Efficiency Analyzer (Handy PEA, Hansatech Instruments, King's Lynn, Norfolk, UK). Bark chlorophyll fluorescence was measured by attaching a collar containing 10 apertures (modified leaf clips) as described in Johnstone et al. [4]. The apertures were 35 mm apart and the branches were third order branches approximately 15 cm in diameter, accessed by an Elevating Work Platform. The apertures were also closed for 30 min until a steady state was achieved.

Once the darkening period was complete, a red light was then flashed (modulated) onto the bark surface with the Handy PEA and an increase in the yield of chlorophyll fluorescence occurred over 1 s (induction curve), until the PS II reaction centers progressively closed [5]. Values for each tree were averaged for the 10 readings. Readings well outside a normal range were eliminated from the data set.

Simple linear regression analysis was undertaken with Minitab 17 software to identify relationships between assessment techniques and between species.

3. Results

Statistical relationships were observed between the February urban visual vitality index and midday leaf water potential ($p < 0.000$, $r^2 = 0.462$, $n = 38$) (Figure 2a) and branch bark F_v/F_m fluorescence and pre-dawn leaf water potential ($p < 0.001$, $r^2 = 0.454$, $n = 22$) (Figure 2b) when all results were pooled. Elms with lower visual vitality indices had decreased water potentials and elms with lower branch bark F_v/F_m fluorescence also had decreased water potentials.

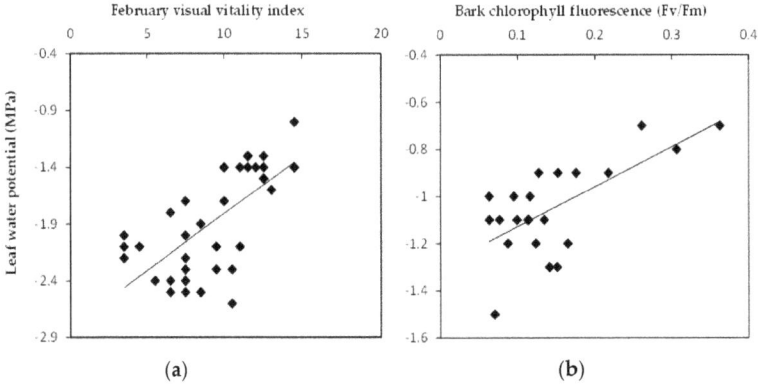

Figure 2. (a) *Ulmus* spp. midday leaf water potential (MPa) versus February Urban Visual Vitality Index ($p < 0.000$, $r^2 = 0.462$, $n = 38$); (b) *Ulmus* spp. pre-dawn leaf water potential (MPa) versus branch bark chlorophyll fluorescence (F_v/F_m) and ($p < 0.001$, $r^2 = 0.454$, $n = 22$).

There were strong individual species differences between the two *Ulmus* species and the measured relationships in this study (Table 1). Statistical relationships were observed between the February urban visual vitality index and midday leaf water potential for *Ulmus* × *hollandica*, but not for *Ulmus procera* (Table 1). Similarly, statistical relationships were observed between branch bark F_v/F_m fluorescence and pre-dawn leaf water potential in *Ulmus* × *hollandica*, but again not for *Ulmus procera* (Table 1).

Table 1. Summarised results from basic statistics and simple linear regression analyses comparing pre-dawn water potentials and midday water potentials with branch bark F_V/F_M in both *Ulmus* × *hollandica* and *Ulmus procera*.

Statistical Value	N	Mean		Minimum		Maximum		SD Value		Regression Statistics	
Comparisons		UVVI	MDWP	UVVI	MDWP	UVVI	MDWP	UVVI	MDWP	p	r^2
U. hollandica; uvvi v mdwp	32	9.7	−1.82	3.5	−2.5	14.5	−1.0	3.225	0.454	<0.000 [a]	0.528
U. procera; uvvi v mdwp	6	7.3	−2.07	6.5	−2.5	8.5	−1.7	0.753	0.350	0.629	0.064
Ulmus data; uvvi v mdwp	38	9.3	−1.87	3.5	−2.6	14.5	−1.0	3.092	0.458	<0.000 [a]	0.462
Comparisons		BB F_V/F_M	PDWP	BB F_V/F_M	PDWP	BB F_V/F_M	PDWP	BB F_V/F_M	PDWP		
U. hollandica; bb F_V/F_M v pdwp	16	0.149	−1.03	0.063	−1.5	0.363	−0.7	0.091	0.212	<0.001 [a]	0.569
U. procera; bb F_V/F_M v pdwp	6	0.138	−1.10	0.095	−1.3	0.175	−0.9	0.031	0.155	0.822	0.014
Ulmus data; bb F_V/F_M v pdwp	22	0.146	−1.05	0.061	−1.5	0.363	−0.7	0.078	0.197	<0.001 [a]	0.454

[a] Statistical relationship is significant and positive; N = number of samples; UVVI = urban visual vitality index; MDWP = midday water potential; BB F_V/F_M = branch bark F_V/F_M; PDWP = pre-dawn water potential; SD value = the standard deviation of a group of variables; p = probability for the t test that the coefficient of the independent variable is equal to zero; r^2 = variation in the dependent variable that can be explained by the urban visual vitality index or branch bark $F_V F_m$ data.

4. Discussion

This study aimed to test practical methods to predict tree stress, particularly long-term moisture stress. The average value for pre-dawn water potential in this study for elms was −1.0 MPa. This indicates that the elms were moderately stressed [14,15]. However, most of the elms had been experiencing moderate to severe drought stress over a significant period, as evidenced by their low urban visual vitality indices. The average score for the elm urban visual vitality index was nine out of a possible 17. The regression analysis of the elm data indicates that the urban visual vitality index used in this study can be used to predict decreased midday leaf water potential in elm populations and in particular within the species *Ulmus × hollandica* (Figure 2a; Table 1). There were, however, no statistical relationships observed between the February urban visual vitality index and midday leaf water potential in *Ulmus procera* when the six trees of this species were analyzed separately (Table 1). This can be explained by the very low variability in the February visual vitality and midday water potential data in *Ulmus procera* (Table 1). It is also a function of the very low number of samples for this species. The pooled Elm results suggest that the urban visual vitality index could be used as an indication of drought stress in mature elms in Summer through visual means and without the need for specialized equipment.

There were no statistical relationships observed between branch bark F_v/F_m fluorescence and pre-dawn leaf water potential in *Ulmus procera* when the six trees of this species were analyzed separately (Table 1). This can be explained by the very low variability visible in branch bark F_v/F_m fluorescence and pre-dawn leaf water potential data in *Ulmus procera* (Table 1). The lack of a relationship could again be a function of the very low number of samples for this species. But again, when all Elm results were pooled or *Ulmus × hollandica* was examined separately, bark chlorophyll fluorescence testing on large branches could predict pre-dawn water status, suggesting that this technique could also be used to assess drought stress in mature elms (Figure 2b; Table 1). Branch bark chlorophyll fluorescence would theoretically enable the year-round testing of deciduous trees, unlike techniques that require leaf samples. Evidence from a previous study on *Eucalyptus globulus* suggests that bark chlorophyll fluorescence may correlate with longer term vitality, whereas leaf chlorophyll fluorescence may be dependent on the health of individual leaves [4].

Under existing and predicted impacts of climate change, it is likely that globally, trees within many urban environments will be subject to altered growing conditions [16]. The ability of trees to adapt to these changing environments will vary; however, trees that have low vitality or that are nearing the end of their lifespan are likely to be less tolerant to change. Identifying suitable assessment techniques for monitoring the vitality of drought stressed mature urban trees will provide tree managers with additional resources for maintaining valuable mature tree resources.

5. Conclusions

We found relationships between the Elm urban visual vitality index and water status and between Elm branch bark fluorescence and leaf water potential, suggesting that both methods can be effectively used for the assessment of long-term drought stress in mature trees. Branch bark fluorescence may be used as an assessment tool even when trees have no leaves in winter and is not dependent on the health of individual leaves. The urban visual vitality index and branch bark fluorescence can aid in the assessment of canopy cover—an essential part of environmental service provision.

Author Contributions: D.C., P.M., and D.M.J. conceived and designed the experiments; D.C. and D.M.J. performed the experiments; D.C. and D.M.J. analyzed the data; D.C. and D.M.J. wrote the paper.

Acknowledgments: The in-kind support of the City of Melbourne, Australia during this research is gratefully acknowledged. The authors also wish to thank Patricio Sepulveda for assistance with field work, and Gerd Bossinger and Ian Woodrow for their preliminary review of this manuscript.

Conflicts of Interest: The authors declare no conflict of interest. The founding sponsors had no role in the design of the study; in the collection, analyses, or interpretation of data; in the writing of the manuscript, and in the decision to publish the results.

References

1. May, P.; Livesley, S.; Shears, I. Managing and monitoring tree health and soil water status during extreme drought in Melbourne, Victoria. *Arboricult. Urban For.* **2013**, *39*, 136–145.
2. Johnstone, D.; Moore, G.; Tausz, M.; Nicolas, M. The measurement of plant vitality in landscape trees. *Arboricult. J.* **2013**, *35*, 18–17. [CrossRef]
3. Percival, G. The use of chlorophyll fluorescence to identify chemical and environmental stress in leaf tissue of three oak (*Quercus*) species. *J. Arboricult.* **2005**, *31*, 215–227.
4. Johnstone, D.; Tausz, M.; Moore, G.; Nicolas, M. Chlorophyll fluorescence of the trunk rather than leaves indicates visual vitality in *Eucalyptus saligna*. *Trees-Struct. Funct.* **2012**, *26*, 1565–1576. [CrossRef]
5. Maxwell, K.; Johnson, G.N. Chlorophyll fluorescence—A practical guide. *J. Exp. Bot.* **2000**, *51*, 659–668. [CrossRef] [PubMed]
6. Percival, G.; Sheriffs, C. Identification of drought-tolerant woody perennials using chlorophyll fluorescence. *J. Arboricult.* **2002**, *28*, 215–223.
7. Grimes, R. *Crown Assessment of Natural Spotted Gum (Eucalyptus maculata), Ironbark (Eucalyptus fibrosa, Eucalyptus drepanophylla) Forest*; Dept. of Forestry: Brisbane, Australia, 1978.
8. Martin, R.A.U.; Burgman, M.A.; Minchin, P.R. Spatial analysis of eucalypt dieback at Coranderrk, Australia. *Appl. Veg. Sci.* **2001**, *4*, 257–266. [CrossRef]
9. Fite, K. Impacts of Root Invigoration (tm) and Its Individual Components on the Performance of Red Maple (Acer rubrum). Ph.D. Thesis, Clemson University, Clemson, SC, USA, 2008.
10. Martinez-Trinidad, T.; Watson, W.T.; Arnold, M.A.; Lombardini, L.; Appel, D.N. Comparing various techniques to measure tree vitality of live oaks. *Urban For. Urban Green.* **2010**, *9*, 199–203. [CrossRef]
11. Bureau of Metrology Commonwealth of Australia. Available online: http://www.bom.gov.au/ (accessed on 25 April 2018).
12. Lindenmayer, D.B.; Norton, T.W.; Tanton, M.T. Differences between wildfire and clearfelling on the structure of montane ash forests of Victoria and their implications for fauna dependent on tree hollows. *Aust. For.* **1990**, *53*, 61–68. [CrossRef]
13. Scholander, P.F.; Hammel, H.; Hemmingsen, E.; Bradstreet, E. Hydrostatic pressure and osmotic potential in leaves of mangroves and some other plants. *Proc. Natl. Acad. Sci. USA* **1964**, *52*, 119–125. [CrossRef] [PubMed]
14. Ranney, T.G.; Whitlow, T.H.; Bassuk, N.L. Response of five temperate deciduous tree species to water stress. *Tree Physiol.* **1990**, *6*, 439–448. [CrossRef] [PubMed]
15. Walters, M.B.; Reich, P.B. Response of *Ulmus americana* seedlings to varying nitrogen and water status. 1 photosynthesis and growth. *Tree Physiol.* **1989**, *5*, 159–172. [CrossRef] [PubMed]
16. Allen, C.; Macalady, A.; Chenchouni, H.; Bachelet, D.; McDowell, N.; Vennetier, M.; Kitzberger, T.; Rigling, A.; Breshears, D.; Hogg, E. A global overview of drought and heat-induced tree mortality reveals emerging climate change risks for forests. *For. Ecol. Manag.* **2010**, *259*, 660–684. [CrossRef]

© 2018 by the authors. Licensee MDPI, Basel, Switzerland. This article is an open access article distributed under the terms and conditions of the Creative Commons Attribution (CC BY) license (http://creativecommons.org/licenses/by/4.0/).

Article

Spatio-Temporal Patterns of Urban Forest Basal Area under China's Rapid Urban Expansion and Greening: Implications for Urban Green Infrastructure Management

Zhibin Ren, Xingyuan He, Haifeng Zheng and Hongxu Wei *

Key Laboratory of Wetland Ecology and Environment, Northeast Institute of Geography and Agroecology, Chinese Academy of Sciences, Changchun 130102, China; renzhibin1985@163.com (Z.R.); hexingyuan@iga.ac.cn (X.H.); zhenghaifeng@iga.ac.cn (H.Z.)
* Correspondence: weihongxu@iga.ac.cn; Tel.: +86-431-8253-6084; Fax: +86-431-8253-6084

Received: 13 April 2018; Accepted: 11 May 2018; Published: 17 May 2018

Abstract: Urban forest (UF) basal area is an important parameter of UF structures, which can influence the functions of the UF ecosystem. However, the spatio-temporal pattern of the basal area in a given UF in regions under rapid urbanization and greening is still not well documented. Our study explores the potential of estimating spatio-temporal UF basal area by using Thematic Mapper (TM) imagery. In our study, the predicting model was established to produce spatiotemporal maps of the urban forest basal area index in Changchun, China for the years 1984, 1995, 2005, and 2014. Our results showed that urban forests became more and more fragmented due to rapid urbanization from 1984 to 1995. Along with rapid urban greening after 1995, urban forest patches became larger and larger, creating a more homogeneous landscape. Urban forest and its basal area in the whole study area increased gradually from 1984 to 2014, especially in the outer belts of the city with urban sprawl. UF basal area was 27.3 × 10^3 m^2, 41.3 × 10^3 m^2, 45.8 × 10^3 m^2, and 65.1 × 10^3 m^2 of the entire study area for the year 1984, 1995, 2005, and 2014, respectively. The class distribution of the UF basal area index was skewed toward low values across all four years. In contrast, the frequency of a higher UF basal area index increased gradually from 1984 to 2014. Besides, the UF basal area index showed a decreasing trend along the gradient from suburban areas to urban center areas. Our results demonstrate the capability of TM remote sensing for understanding spatio-temporal changing patterns of UF basal area under China's rapid urban expansion and greening.

Keywords: urban forest; urbanization; sampling plots; Landsat TM; basal area

1. Introduction

During the last three decades, many serious urban environmental problems have evoked considerable social concerns in China [1]. Many cities in China have set up a lot of environmental improvement strategies such as a focus on urban forest (UF) establishment [2]. UFs in cities are the most important parts of urban ecosystems [3]. The establishment of UFs can be considered as an important way to improve the urban environment [3]. Urban forests could provide many ecological functions to resolve urban environmental problems [4,5], such as reducing urban air pollutant concentrations [6,7], sequestering atmospheric CO_2 [8,9], reducing storm water runoff [10,11], mitigating the urban heat island [12,13], and providing a habitat for organisms [14,15].

UF basal area is considered as an important component of UF structures, which can influence UF ecosystem functions [16,17]. In the past six decades, China has experienced rapid urbanization and urban greening [18]. The urban population in China is predicted to reach 1.5 billion with an urbanization level of 50% by the end of 2020 [19]. With the continuous development of urbanization,

UF has great development potential in China [20]. The Chinese government have placed an increasing emphasis on UF development in recent decades [21,22]. UF City Programs were proposed by the Chinese government from 2004. Many governmental regulations relating to urban greening have been introduced. Urban forests in China could experience a dramatic change due to intensive human activities, such as urbanization and urban greening. However, the comprehensive effects of urbanization and greening on the spatio-temporal patterns of UF basal area have still not been understood. Therefore, the accurate and timely estimation of spatiotemporal UF basal area is necessary and useful for urban managers to understand UF functions and maximize the environmental benefits of UFs under China's rapid urban expansion and greening. To the best of our our knowledge, the spatiotemporal dynamic patterns of the UF basal area under China's rapid urbanization and greening have rarely been studied and are not yet fully understood.

As we know, the acquisition of spatial-temporal UF basal area often depends on conventional intensive and costly plot-based field work [23–25]. In addition, it is very tough to obtain spatial-temporal patterns of UF basal area at the urban landscape scale through direct field sampling measurements. The lack of a consistent area-wide UF basal area would impact our ability to conduct the ecological analyses of UF functions at a landscape level. Using remote sensing techniques, the estimation of UF basal area should be connected with various vegetation indices developed from remote sensing data to overcome these limitations [26]. Among these remote sensing data, the higher resolution remote sensing dates such as QiuckBird remote sensing and Systeme Probatoire d'Observation dela Tarre (SPOT) or eye-level photography can be used for extracting urban forest structures with a high accuracy. However, these data are often more expensive. While Landsat Thematic Mapper (TM or Enhanced Thematic Mapper (ETM+)) imagery has a poorer resolution and issues with mixed pixels, it is easily accessed, less expensive, and widely used all over the world to estimate forest structural attributes. Many researchers have demonstrated that some indices obtained from TM or ETM$^+$ imagery data such as the Normalized Difference Vegetation Index (NDVI), simple ratio (SR), and green normalized difference vegetation (GNDVI), are significantly correlated with forest structural attributes measured on the ground such as canopy cover, stem density, diameter at breast height, tree height, base area, leaf area index, biomass, etc. [26–30]. The most commonly used spectral indicator extracted from TM or ETM$^+$ imagery is the Normalized Difference Vegetation Index (NDVI). Some researchers have showed that the NDVI has a significant relationship with ground measured natural forest structures [28,29]. Previous studies have achieved some degrees of success in estimating the forest basal area index from TM or ETM$^+$ data in natural areas worldwide. However, the conclusions about relationships between forest basal area and NDVI vary, depending on the characteristics of the study areas [30]. UFs are usually very different from natural forests, which are heterogeneous, fragmented and scattered, and surrounded by many impervious surfaces [16,17]. Therefore, the relationships found between vegetation indices and natural forest structures may be different from the relationships between vegetation indices and urban forest structural attributes. Whether NDVI extracted from TM or ETM$^+$ imagery can still be used for estimating UF basal area is still unknown. Based on our literature review [23–29], there have been few studies on estimating spatiotemporal patterns of UF basal area with TM or ETM$^+$ imagery.

Based on field measurements and TM remote sensing data acquired in four different years (1984, 1995, 2005, and 2014) from the City of Changchun, China, our study aims to characterize the changing patterns of UF basal area from Landsat TM imagery from 1984 to 2014 under two key forces: the fast urbanization process and the urban greening policies. The purposes of our research are to: (1) examine the usefulness of TM remote sensing at different times in estimating spatio-temporal changes of UF basal area; (2) develop a model for predicting UF basal area by coupling field measurements with TM remote sensing; (3) explore the dynamic spatio-temporal patterns of UF basal area in the City of Changchun, China from 1984 to 2014; and (4) study the implications of UF establishments for urban environmental improvement under China's rapid urban expansion and greening.

2. Methods

2.1. Study Area

Our study was conducted within the fifth-ring road of Changchun (125°09′–125°48′ E, 43°46′–43°58′ N) (Figure 1), which is the capital of Jilin Province and an important social-economic center of northeastern China. Changchun is located in the hinterland of the Northeast Plain and had a total population of 3.6 million by the end of 2010. The average total yearly precipitation in Changchun is 567 mm and the average temperatures of the cold winter and hot summer are −14 °C and 24 °C, respectively. Changchun is called the "Forest city", with an average forest cover rate of 45%. Urban forest species are the most abundant with 43 families, 86 genera, and 211 species [24]. Since 1980, Changchun has experienced an accelerated progress of urbanization in temperate regions of Northern China [31], which might have caused dramatic changes of UF structure and species composition. Therefore, Changchun is an ideal city for analyzing the spatiotemporal patterns of UF basal area under China's rapid urban expansion and greening.

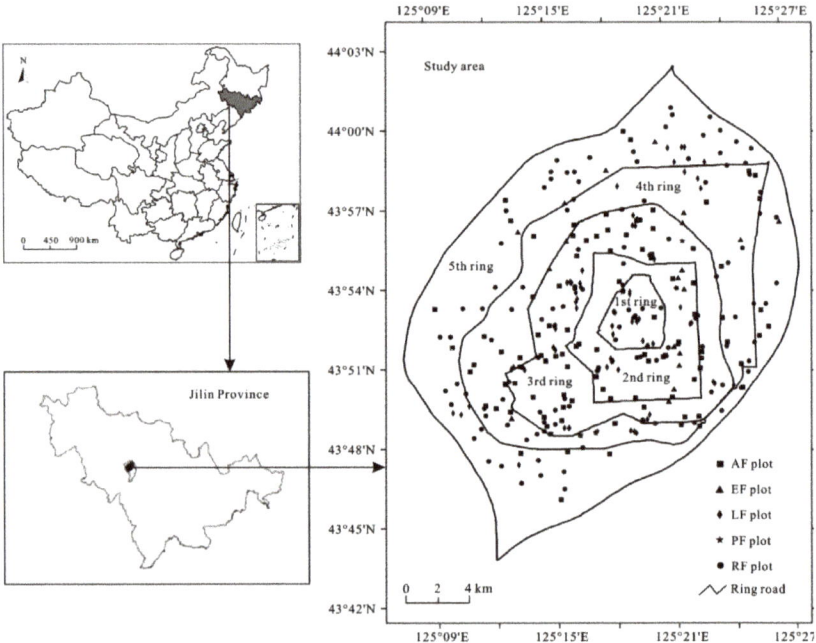

Figure 1. The study area located within the fifth-loop road in the City of Changchun, China. AF: attached forest; EF: ecological and public welfare forest; LF: landscape and relaxation forest; PF: production and management forest; RF: road forest.

2.2. Image Data and Processing

The four scenes of TM images with a resolution of 30 m were collected on 14 September 1984, 29 September 1995, 8 September 2005, and 3 October 2014 with a cloud cover less than 5% of scenes. These four Landsat scenes across three decades were within the same phenological stage. In Changchun, the trees began shedding their leaves in early November and the leaves were still on the trees on 3 October. Therefore, the four scenes of TM images can be used in our study for urban forest research. The atmospheric correction for the TM images was first undertaken and then the TM raw digital numbers (DN) were converted into surface radiance values by following the procedures provided by Chander and Markham [32]. Finally, the TM images were geo-referenced to the Universal Transverse Mercator (UTM) coordinate system with a root mean square error (RMSE) of less than 0.5 pixel by using 33 ground control points taken from topographic maps. Based on four scenes of TM images, the Normalized Difference Vegetation Index (NDVI) index was further calculated in ENVI 4.6 through the equation of NDVI = (b4 − b3)/(b4 + b3), where b3 and b4 are the surface reflectance values in TM bands 3 and 4, respectively. To conduct a spatiotemporal analysis of UF basal area with multitemporal TM images, it is necessary to normalize NDVI maps calculated from the multitemporal TM images to eliminate environmentally introduced radiometric effects. The relative radiometric correction method of pseudo-invariant features (PIF) was applied in our study [33,34]. This procedure uses one image as a reference image (2014) and adjusts the radiometric properties of all other images (1984, 1995, and 2005) to it by the analysis of invariant features, such as roads, rooftops, and deep water. In our study, seventy-five spatial evenly distributed regions of interest for invariant features (including 25 from roads, 30 from rooftops, and 20 from water bodies) located on the mul-titemporal images were manually selected. The average NDVI in each interest region was then used to develop a linear normalization model between the reference image (2014) and the subject (i.e., uncorrected) images (1984, 1995, and 2005). The normalized subject images (Figure 2) were obtained using the following equation:

$$NDVI_{ref} = a\ NDVI_{sub} + b$$

where $NDVI_{ref}$ is the reference image (i.e., 2014 image); and $NDVI_{sub}$ is the subject image (i.e., 1984, 1995, and 2005 images). The scene normalization coefficients of NDVI before and after normalization are listed in Table 1.

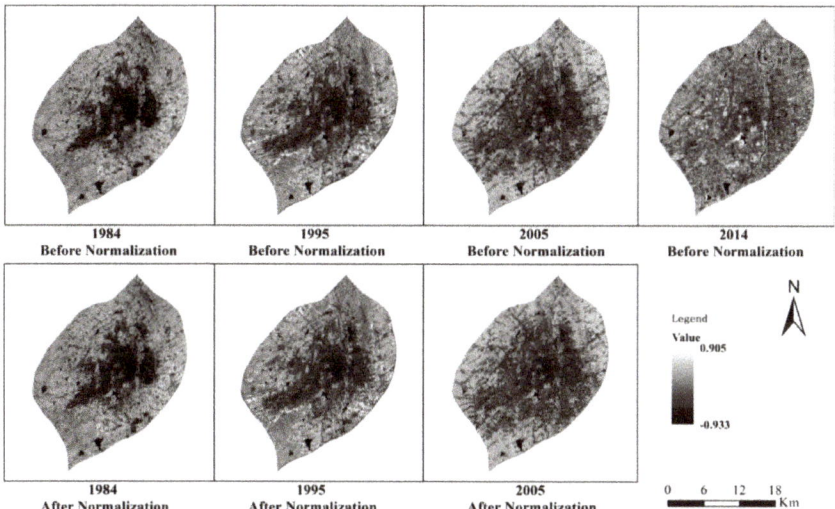

Figure 2. The NDVI images before and after normalization.

Table 1. The image normalization coefficients and the correction coefficients for each subject image.

	1984	1995	2005
a	0.712 **	0.745 **	0.876 **
b	−0.154 **	−0.103 **	−0.095 **
R^2	0.725	0.712	0.842
p-value	0.00	0.00	0.00

** Significant at the 0.01 probability level.

2.3. Sampling Design and UF Basal Area Calculation

Plot measurements: 159 random plots (sampling plots) throughout the study area were established in this study by using the methods from the Urban Forest Effects (UFORE) Model [35]. Field measurements of UFs were conducted during July and August 2013 and 2014 (Figure 3). The sampling plots were randomly selected to ensure that they were representative of the major types of urban landcover such as residential areas, road areas, park areas, and commercial areas in Changchun. Besides, sampling plots were required to be located in relatively homogenous patches greater than 1600 m^2, and in this study, each of the 159 sampling plots was defined as a 30 m × 30 m (0.09 ha) area to represent a TM pixel size. The coordinates of each sampling plot were recorded with a global positioning system (MG838GPS, UniStrong Company, Beijing, China) with the high-accuracy better than 1 m, which were used to extract the NDVI value from multi-temporal TM derived NDVI maps. A total of 5693 tree individuals were measured from the 159 sampling plots. At each sampling plot, some UF structural attributes including tree species, vegetation types, stem density, diameter at breast height (DBH), tree height (H), and crown size, were measured or collected. Finally, UF basal area (m^2) was considered as the cross-sectional area of all trees in a sampling plot and calculated at each plot in this study. The basal area index (in m^2/ha) is defined as the ratio of the cross-sectional area of all trees in a sampling plot to the plot ground area, as shown in Equation (1).

$$\text{Basal Area index } (m^2/ha) = \frac{\left(\sum_{i}^{N} \pi (DBH_i/2)^2\right)}{0.09} \quad (1)$$

where N is the number of trees in a sampling plot for the equation.

After the calculation of UF basal area, NDVI values were extracted from the normalized TM images in ArcGIS 9.3 software (Environmental Systems Research Institute, Redlands, CA, USA) with the latitude and longitude of each sampling plot for later statistical analyses.

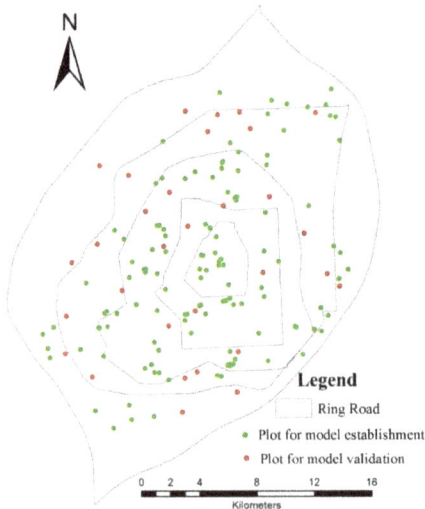

Figure 3. Map of 159 sampling plots in the City of Changchun.

2.4. Spatio-Temporal Patterns of UF and Its Basal Area with TM Images

UF can be defined as a synthesis between an organism and abiotic environment, which should reach a big enough area (>0.5 ha) and coverage with trees as the main body, considerably influence the surrounding environment, and provide ecological values and human landscape values [36]. In our study, it should be noted that only patches of urban vegetation with an area larger than 0.5 ha were extracted as UF. Based on four scenes of TM images, an object-based approach was used for UF extraction [37]. This approach did not classify single pixels, but groups of pixels that represent already existing objects in ArcGIS software (Environmental Systems Research Institute, Redlands, CA, USA), which is based on a supervised maximum likelihood classification. The TM images were firstly segmented into objects by an object-based approach and then assigned to a UFs class. An object-based approach is superior to traditional pixel-based classification [38]. For example, the salt-and-pepper noise is a form of noise sometimes seen on images, which can be caused by sharp and sudden disturbances from the image signal. It presents itself as sparsely occurring white and black pixels. The "salt-and-pepper" effect frequently found in pixel-based classification can be largely avoided when using an object-based approach for land cover classification. The historical high spatial resolution images in Google Earth were used as reference data for the accuracy assessment. An accuracy estimate was conducted based on 200 checkpoints for every respective year with historical high spatial resolution images in Google Earth as references for 1984, 1995, 2005, and 2014. The overall accuracies of UF extraction for 1984, 1995, 2005, and 2014 were 89.24%, 90.43%, 91.58%, and 93.12%, respectively.

In order to produce the spatiotemporal maps of UF basal area index from historical TM imagery, NDVI was used to build the prediction model for UF basal area index. The corresponding regression model between UF basal area index collected from the 129 plots and NDVI extracted from the 2014 TM image was established to uncover quantitative relationships between them. In the analyses, the plot-based NDVI was used as the independent variable, while UF basal area index was used as a dependent variable. In this study, coefficient of determination (R^2) for regression analysis between NDVI and UF basal area index was calculated to assess the relationship. To evaluate the reliability and accuracy of the established models, plot-based measured UF basal area index data at 30 plots (Figure 3) were used for validation. The variability is represented by the standard error. R^2 and RMSE were used to test the fitness of the predicting model at plots. A well-calibrated model should have a root mean square error (RMSE) that is small relative to the total observed variation and an R^2 close to one. All statistical analyses were carried out with standard statistical software, SPSS (Version 19.0, IBM Company., Chicago, IL, USA).

Mapping UF basal area index: a vector layer data of UF was first used to extract the NDVI of UF from 1984, 1995, 2005, and 2014 TM images in our study, and then the map of UF basal area index in 2014 was created by calculating pixel-based values of UF basal area index using the regression model developed with NDVI extracted from the 2014 NDVI image at the 129 plots and 2013–2014 field survey data. We also created spatio-temporal maps of UF basal area index from normalized NDVI images calculated from 1984, 1995, and 2005 TM images using the regression model of UF basal area index, developed with the 2014 NDVI image.

3. Results

3.1. Urbanization in Changchun from 1980 to 2014

Some variables of urbanization in Changchun were collected from the Statistics Yearbook of Cities in China (National Bureau of Statistics of China, 1980–2014). These variables include urban population, built-up area, and gross domestic product (GDP) per capita, which were used to assess the urbanization level in our study. The results showed that Changchun has experienced rapid urbanization during the last 30 years (1984–2014) (Figure 4). Compared to 89 km^2 in 1984, urban build-up area reached 430 km^2 in 2014, with a rapid increment during last three decades. Urban population increased from 110×10^4 in 1984 to 410×10^4 in 2014, with an increment of 9×10^4 for each year. Meanwhile, GDP area per Capita also showed a rapid increase from 1984 to 2014.

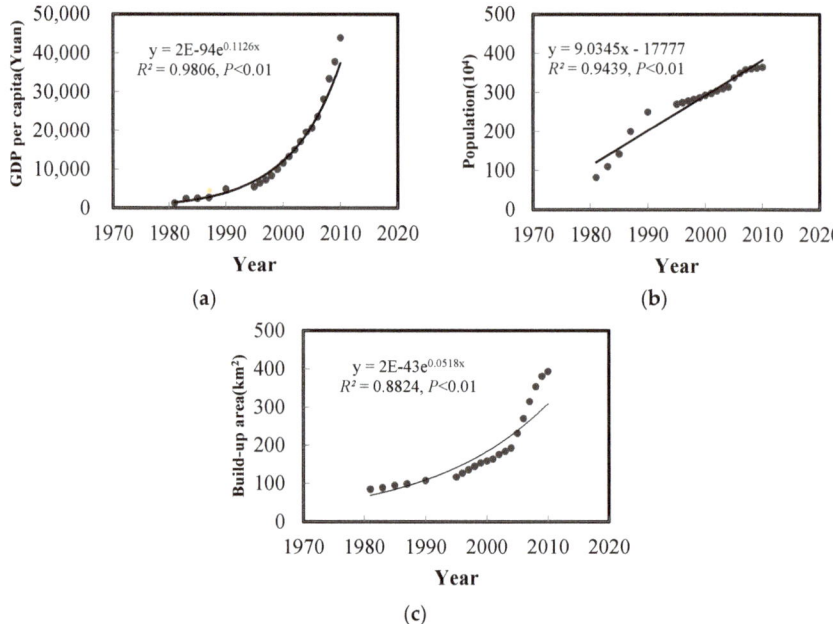

Figure 4. Urbanization process in Changchun city from 1980 to 2014. (**a**) GDP; (**b**) Population; (**c**) Build-up area.

3.2. Spatial Assessment Model for UF Basal Area Index

In order to produce the spatio-temporal maps of UF basal area index from historical TM imagery, NDVI was used to build the prediction model for UF basal area index. The results showed that the UF basal area index had a positive non-linear relationship with NDVI (Figure 5), which suggested that the UF basal area index increases non-linearly with the increase of NDVI. As seen in Figure 5, when NDVI was lower than 0.5, the UF basal area index increased slowly with the increase of NDVI. However, when NDVI was larger than 0.5, the UF basal area index increased sharply (Figure 5). The non-linear model with NDVI as the independent variable could explain 69.2% of total variance of UF basal area index. Finally, the established regression model was then applied to produce the maps of UF basal area index from normalized historical NDVI images in 1984, 1995, 2005, and 2014, respectively. To evaluate the reliability and accuracy of the predicted model, 30 plot-based measured UF basal area index data were used for validation. A well-calibrated model should have a root mean square error (RMSE) that is small relative to the total observed variation and an R^2 close to one. Our results showed that the modeled UF basal area index compared closely with the plot-measured UF basal area index (RMSE = 1.3 m^2/ha; R^2 = 0.90) (Figure 6). Ninety percent of the variance of UF basal area index among the 30 plots can be captured by the model. However, high UF basal area index values were slightly underestimated. Therefore, the established model is accurate and can be used to predict the UF basal area index.

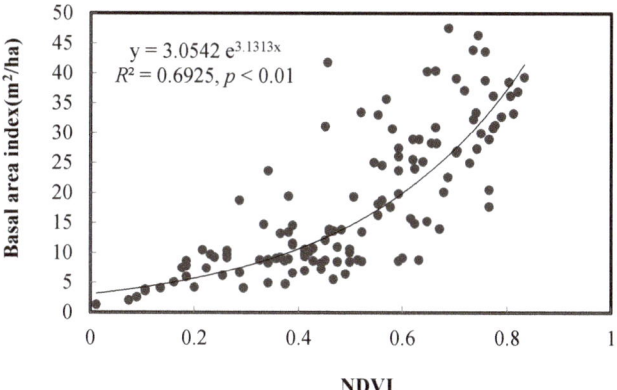

Figure 5. Regression analyses of NDVI with UF basal area index ($n = 129$, $p < 0.01$).

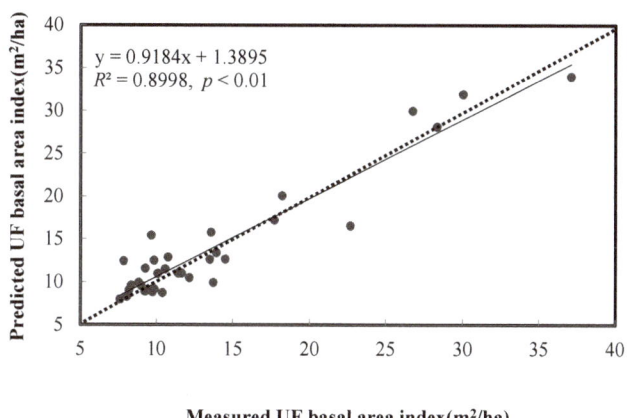

Figure 6. Comparisons between plot-based measured UF basal area index and the modeled UF basal area index at the 30 sites ($n = 30$).

3.3. Spatio-Temporal Distribution of UF and Its Basal Area

The spatio-temporal patterns of UF over the past three decades (1984–2014) have changed dramatically (Figure 7), mirroring the intense urbanization and rapid development of urban greening in China. It is obvious that UF gradually expanded from downtown to the suburban area during the study period. UF coverage steadily increased from 15% in 1984 to 25% in 2014 (Table 2). The UF area in 2014 increased by 63.1% (52.4 km^2) compared to 1984. Our results revealed the clear changes of UF spatial configuration. The total number of UF patches in the Municipality area approximately increased by 99% from 1984 to 1995, and then decreased by 2.6% from 1995 to 2005 and 27.8% from 1995 to 2014 (Table 2). Correspondingly, the values of mean UF patch size showed a decline of 43.4% from 1984 to 1995, but exhibited an increase of 14.9% from 1995 to 2005, and the increase was doubled from 2005 to 2014. These changes in landscape configuration suggest that the UF became increasingly fragmented due to intense urbanization from 1984 to 1995. Along with rapid urban greening after 1995 (Figures 7 and 8, Table 2), the UF patches became larger and larger, creating a more homogeneous landscape when compared with the spatial patterns in 1995. UF was mostly located in the urban central area in 1984 (Figures 7 and 8) and then distributed more evenly across the whole urban area in more recent years (2005 and 2014). In the study years of 1984, 1995, and 2005 (Figures 7 and 8), the UF coverage was higher in the two-ring area than that in the other-ring area. In 2014, the highest value of UF coverage occurred in the three/four-ring area. Spatial heterogeneity of UF change also existed from 1984 to 2014. Our results showed that UF coverage in the suburban area (four to five-ring area) increased gradually from 1984 to 2014, but decreased sharply in the urban central area (one/ two-ring area) of the city (Figures 7 and 8). UF Patch numbers showed a decreasing trend from suburban areas to downtown for all the study years (Figures 7 and 8). From 1984 to 2014, UF patch numbers increased in all ring areas, especially in the five-ring area. Mean patch area of UF showed an increasing trend from suburban areas to the downtown area for the years 1995, 2005, and 2014 (Figures 7 and 8). Mean patch area of UF in the suburban area (five-ring) increased gradually from 1984 to 2014, but decreased sharply in the downtown area (one/two-ring area) of the city.

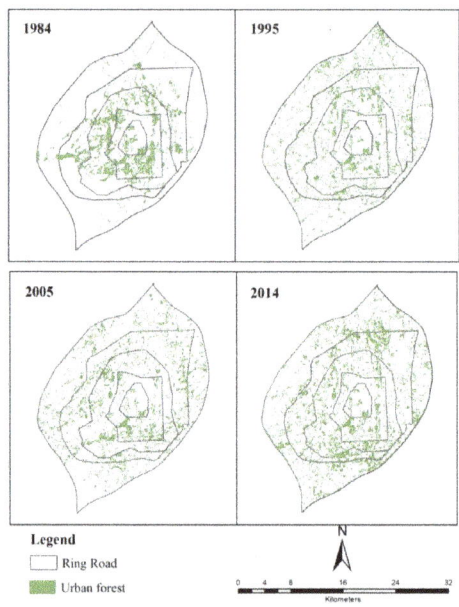

Figure 7. Spatio-temporal patterns of UF in Changchun during 1984–2014.

Table 2. Summary of UF attributes within the study area in the City of Changchun, China.

Year	Total UF Area (10^3 m^2)	Coverage (%)	Patch Number (n)	Mean Patch Area (10^3 m^2)	Patch Density (n/km^2)	Basal Area (m^2)	BA Index (m^2/ha)
1984	78,585.612	0.152	1014	77.501	12.912	27,294.725	0.521
1995	89,084.842	0.174	3013	29.621	33.842	41,338.945	0.793
2005	104,800.523	0.242	2934	35.719	27.934	45,760.852	0.874
2014	133,564.224	0.251	2174	61.337	16.325	65,139.626	1.245

Note: UF basal area (m^2) was considered as the cross-sectional area of all trees in this study area. Basal area index (m^2/ha) = Total basal area/study area.

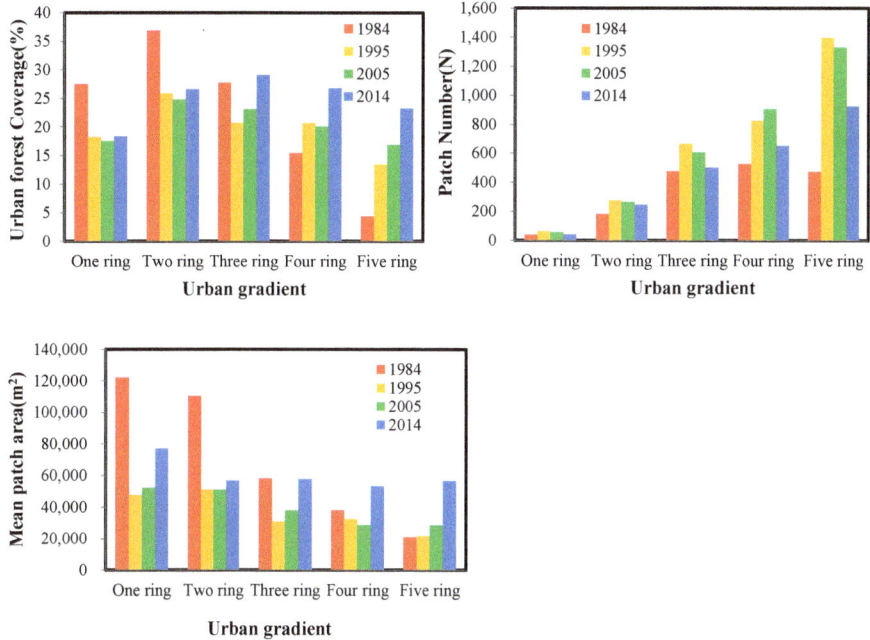

Figure 8. Urban forest coverage, patch characteristics, and basal area across ring road-based urban gradient.

By analyzing the pixel-based maps of UF basal area index in a GIS format, we found that they were highly dynamic among the years 1984, 1995, 2005, and 2014 (Figures 9 and 10). UF basal area was 27.3×10^3 m^2, 41.3×10^3 m^2, 45.8×10^3 m^2, and 65.1×10^3 m^2 of the entire study area in 1984, 1995, 2005, and 2014, respectively. UF basal area in Changchun increased gradually from 1984, 1995, and 2005 to 2014, mirroring the intense urbanization and rapid development of urban greening in the city. In addition, the high spatial heterogeneity of the basal area was observed across the city. In 1984, pixels with values of UF basal area were mostly concentrated in downtown areas (Figures 9 and 10) and then distributed more evenly across the whole urban area in more recent years (2005 and 2014). UF basal area showed a decreasing trend from suburban areas to downtown areas in the years 1995, 2005, and 2014 (Figure 10). Meanwhile, the UF basal area index was higher in the two-ring area than that in the other-ring area in the city of Changchun for the years 1984, 1995, 2005, and 2014 (Figure 10). Our results also show that the different spatial changes of UF basal area have occurred across different urban gradients from 1984 to 2014. The UF basal area increased more in the suburban area than that in the downtown area (Figure 10). However, the UF basal area index exhibited a greater increase in the downtown area than that in the suburban area.

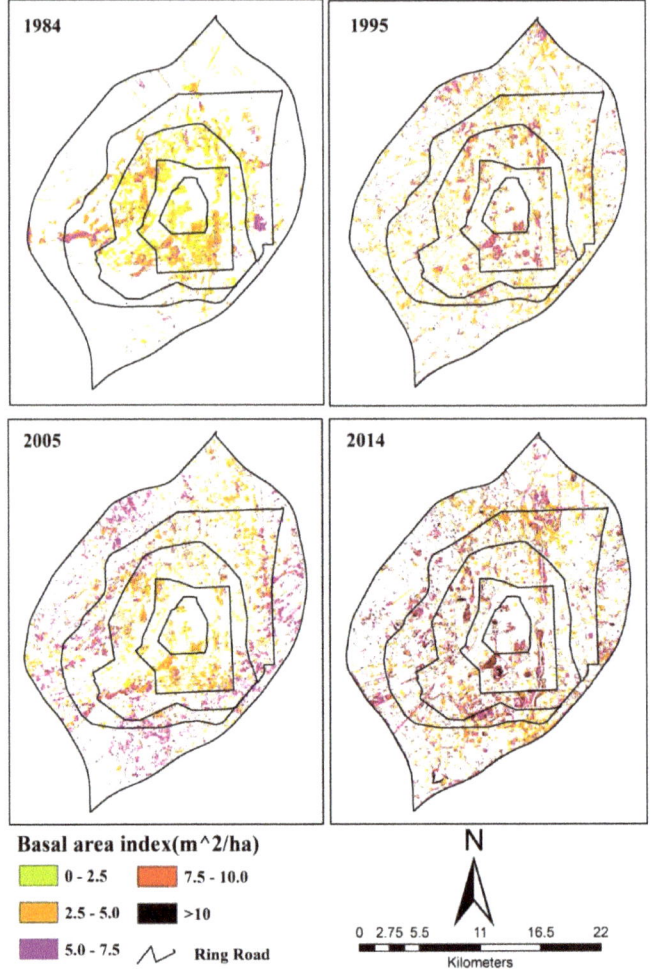

Figure 9. Spatiotemporal distribution of UF basal area index in the City of Changchun.

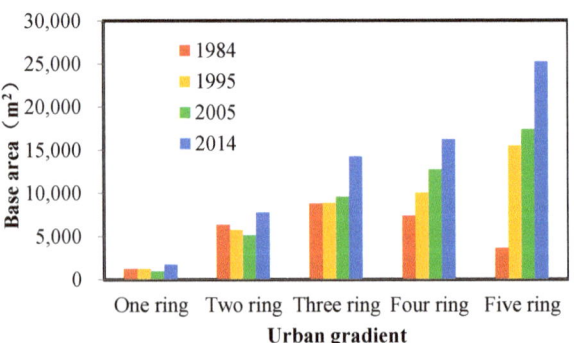

Figure 10. *Cont.*

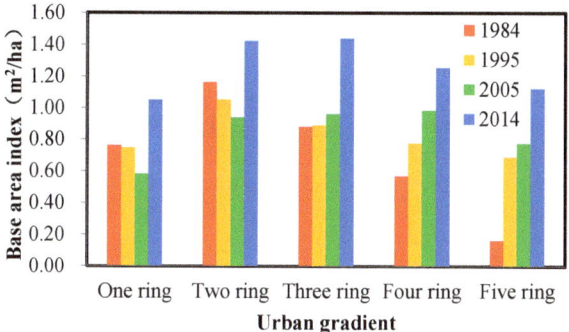

Figure 10. UF basal area across urban gradient: UF basal area (m^2) = the cross-sectional area of all trees in the study area. Basal area index (m^2/ha) = Total basal area/study area.

The class distributions of UF basal area index were all skewed toward low values in 1984, 1995, 2005, and 2014. The results showed that the UF basal area index in 1984, 1995, 2005, and 2014 with the highest frequency was 2.5–5 m^2/ha (Figure 11). The frequency of UF basal area index from 2.5–5 m^2/ha was 59%, 71%, 56%, and 45% in 1984, 1995, 2005, and 2014, respectively. However, the frequency of a lower UF basal area index (2.5–5 m^2/ha) decreased gradually from 1984, 1995, and 2005 to 2014. Meanwhile, the frequency of a higher UF basal area index (>5 m^2/ha) increased gradually from 14% in 1997 to 48% in 2014. About 14% of UF basal area index values were above 7.5 m^2/ha in 2014, but there were just a few of pixels with a UF basal area index >7.5 m^2/ha in 1984, 1995, and 2005.

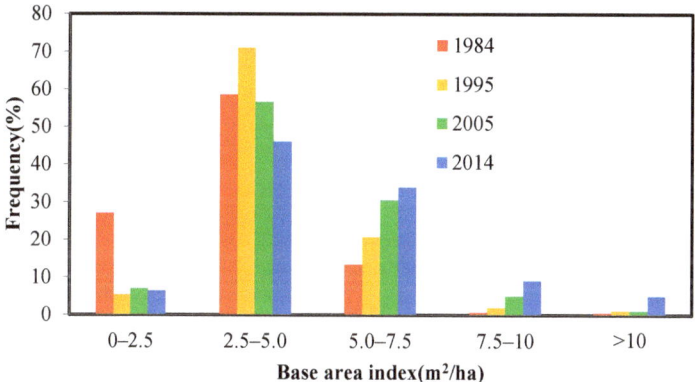

Figure 11. Histograms of the frequency statistics for UF basal area index, calculated from pixels with UF in the study area of Changchun, China.

4. Discussion

4.1. The Spatio-Temporal Estimation of UF Basal Area Index

Our results demonstrated the potential of using mul-titemporal TM imagery to characterize spatiotemporal changing patterns of UF basal area in practice. UF basal area could be still estimated by NDVI extracted from TM imagery just like natural forest basal area that could be predicted well by NDVI [27,30]. However, the relationship between forest basal area and NDVI was weaker in the urban area than that in the natural area [8,28]. One possible reason for this might be that the environment of UFs is very different from that of natural forests. Usually, urban heterogeneous and fragmented environments may result in a mixed pixel problem. Therefore, the mixed pixel issue might be the main reason for the confusion of estimating UF basal area by NDVI. The approach of establishing a model to predict UF basal area has been successful, but has serious limitations. The major limitations of using vegetation indices (VIs) to estimate forest basal area are that VIs frequently lose sensitivity and saturate at a moderately high above-ground biomass (AGB) or leaf area index (LAI). Many studies [39–41] have reported that models between VIs and basal area or biomass are curvilinear and there is a trend of saturation in the VIs [42]. The nonlinear equations between NDVI and UF basal area or biomass found in this study also implied this limitation. Our results show that the saturation effect of the NDVI still existed in UFs similar to that in natural vegetations. However, such a kind of limitation could be ignored because UF basal area values from most urban vegetated-areas in this study were much lower than that in natural forest areas and few of the UFs had a canopy density above 60%. In addition, it should be noticed that there are some limitations of our study. The different climatic conditions, specifically precipitation and temperature, may also influence relationships between NDVI and urban vegetation structural attributes. This should be taken into account when applying the method developed in this study to other cities or at different seasons. More research on the relationships between vegetation indices and urban forest structural attributes for different seasons and cities should be conducted in the future.

4.2. Spatio-Temporal Changes of UF Basal Area under China's Rapid Urbanization and Urban Greening

In this study, we found that UF increased gradually from 1984, 1995, and 2005 and to 2014 (Table 2 and Figures 7 and 8). The UF changing patterns in the metropolitan area of Changchun, China from 1984 to 2014 responded to combined effects of rapid urbanization and greening policies (Figure 7). UF in China rapidly developed, especially in suburban areas, with the accelerating urban sprawl. This is why we found the gradual increases of UF from 1984 to 2014. The opposite phenomenon was reported in other regions, like most Eastern European cities and the USA [43–46]. This difference might be caused by different urban development patterns in different countries. There are two main patterns of urban development in the world [43]. The first pattern is to sprawl into the wider countryside for urbanized lands and the second is to be densified by urbanization level with development of the existing urbanized area. However, in Europe and the USA, some cities developed by "densifying" through the extension of urban area, which might result in the UF decline. In China, most cities are extending by "sprawl" into the wider countryside. Consequently, more new UFs were established. After having realized the important environmental function of UF in urban ecosystems, local governments in China have set out a series of policies such as "constructing forest cities" (designated by the State Forestry Administration of the People's Republic of China) and invested a large amount of money to introduce green elements into urban central areas in order to resolve such environmental problems [47]. In addition, among these are establishing more new UF parks and community gardens, planting more trees along roads, and especially paying more attention to the establishment of many national UF parks in suburban areas. Therefore, this contributed to the sharp increase of suburban forest UF in the study area from 1984 to 1995. Although UF amount increased gradually with the development of urbanization, UF in central urban areas became more and more fragmented due to intense urbanization from 1984 to 1995. However, along with rapid urban greening in recent years, UF patches became larger and larger, creating a more homogeneous landscape when compared with the spatial patterns in 1995. In spite of the increasing UF amount, some planning strategies are still needed. Conservation or construction plans for the forests in the urban core areas of Changchun are desirable to protect forests from potential loss caused by urbanization, particularly in the first-ring road area in the city where built-up land has increased [22].

4.3. Implications for Urban Green Infrastructure Management and Planning

Since 1978 (Chinese reform and opening policies), China has experienced a rapid and unprecedented process of urbanization (Figure 4). For instance, the urban built-up area in Changchun increased by 363.5% from 85 km^2 in the 1980s to 394 km^2 in the 2010s over a 30-year period. With the rapid urbanization, many environmental problems in cities have arisen in China, affecting human health, the quality of urban life, and the sustainability of the urban ecosystem [48]. Urban forests considered as an important part of urban ecosystems have many important ecological effects and make a great contribution to the improvement of urban environments [16,17,49]. The good spatial and landscape planning can help urban patterns to protect ecosystems and thus support the provision of a needed service and solve urban environmental problems [50]. However, the environmental consequences of planning decisions on the landscape are often undervalued. There might be a lack of adequate information about urban forest at the landscape level. With the development of remote sensing technology, the estimation of spatiotemporal basal area by UF from the remotely sensed imagery is very important for us to enhance understanding the dynamics of UF basal area and has important implications for UF development under China's rapid urbanization and urban greening, providing information about how to establish UF to maximize their ecological functions, particularly for cities where UFs are still under construction. In this study, we found that the spatial distribution of UF basal area was very dynamic, mirroring the intense urbanization and rapid development of urban greening in the city. UF basal area increased gradually from 1984 to 2014 (Figures 10 and 12), especially in suburban areas. With the urban expansion, an obvious change of UF basal area first occurred in an urban fringe and suburban area (Figure 12). UF basal area in China underwent a rapid development, especially in suburban areas with the accelerating urban sprawl. Therefore, gradual increases of UF basal area were observed from 1984 to 2014. The UF basal area was also found to be heterogeneous across the different urban gradients (Figure 10). The decreasing trend of UF basal area from suburban areas to downtown areas for all study years is consistent with the decreasing urbanization intensity from downtown areas to suburban areas in China [20]. The changing patterns of UF basal area in the metropolitan area of Changchun, China from 1984 to 2014 responded to combined effects of rapid urbanization and greening policies (Figures 4 and 10).With urban sprawl, UFs developed rapidly, and could serve as important green infrastructure in cities. UF basal area increased from 68×10^3 in 1984 to 224×10^3 in 2014, with an increment of 5.2×10^3 for each year. Meanwhile, our results also showed that most areas in our study area were still covered by the low UF basal area or biomass values. These results may suggest that there is still great potential to increase the capacity of UF basal area and improve urban environments in Changchun. This has some important management implications for urban greening. Firstly, there is still great potential to increase the total area of UFs in Changchun. Currently, the UF cover is only 25%. More trees can be planted and more C can be stored and sequestered. In addition, researchers still need to take some measures for local urban managers to increase UF basal area or biomass. We should seek practical approaches to optimize UF structure to enhance the capacity of these UFs in ecological functions. Some measures could be suggested such as tree species selecting, pruning, and shaping. In practice, native species with fast growth rates should be planted to improve the capacity of carbon sequestration of UFs. Besides, the multilayer forest communities with a high canopy density and LAI are also the most effective in terms of the ecological effect. In our study, our results also show that the UF patterns are uneven in the study area. UF cover in suburban areas was higher than that in urban core areas, especially within the first-loop road in Changchun, which could lead to the environmental inequity [51,52]. Therefore, urban planners and policy makers should be concerned with the distribution inequity of UFs and plant more trees to increase the amount of UFs in urban central areas. In recent years, China has selected ecological civilization as the national strategy to build a beautiful China. It will be a great chance for UF development in China. The government should strengthen the building of UF. UF quality also needs to be further improved, especially in terms of the structure and function for secure good dwelling environments. Our work presented here suggests that the use of UF under China's rapid urbanization and urban greening offers significant potential for urban environmental improvement.

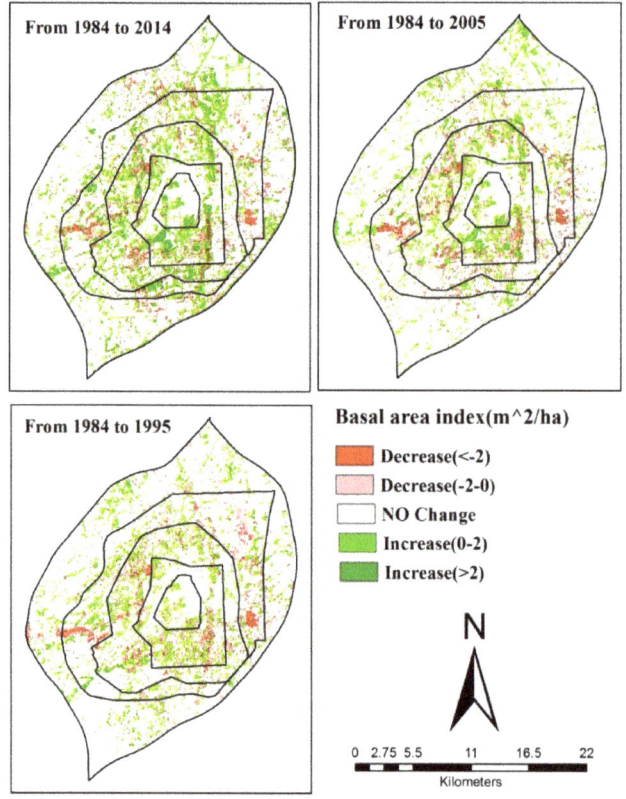

Figure 12. Spatiotemporal changing patterns of UF basal area index in the City of Changchun.

5. Conclusions

Based on multi-temporal Landsat TM data (1984, 1995, 2005, and 2014) and UF field survey data, this study explored the potential of using TM imagery to estimate spatio-temporal patterns of UF and its basal area in the City of Changchun, China. The following conclusions could be realized:

(1) Landsat TM imagery could provide a faster and cost-effective method to obtain spatio-temporal patterns of UF and a 30-m resolution UF basal area. NDVI is still a good predictor to estimate and map UF basal area.
(2) In the rapid urbanized region (within the study area) of Changchun City, the UF and its basal area have been found to increase significantly from 1984, 1995, and 2005 to 2014, especially in the outer belts of the city due to urban sprawl. The UF basal area class distribution was skewed toward low values in 1984, 1995, 2005, and 2014, but gradually skewed toward relatively high values from 1984 to 2014.
(3) The results demonstrate that the spatiotemporal pattern of basal area by UF has great implications for urban forest establishment under China's rapid urbanization and urban greening.

The results from our study provide needed baseline information at the landscape scale by producing the relatively high-resolution maps of UF and its basal area. If this study can be used in practice, more time-saving and labor-saving estimations of spatiotemporal UF basal area could be possible to assess the actual and potential role of UFs to improve the urban environment. Furthermore, the accurate information of UF and its basal area may allow urban planners to conduct more realistic and better planting designs of UFs at the urban landscape scale.

Author Contributions: Zhibin Ren analyzed the data and wrote the paper; Zhibin Ren and Hongxu Wei finished the field investigation; Xingyuan He and Hongxu Wei helped with designing the research and assisted in data analysis and with editing the manuscript; Haifeng Zheng helped with extracting the spatial information of urban forests.

Funding: This research was supported by Funding for Jilin Environmental Science (2017-16), the National Natural Science Foundation of China (41701210), Science Development Project of Jilin Province, China (20180418138FG), the "Strategic Priority Research Program" of the Chinese Academy of Sciences (KFZD-SW-302-03), and the Foundation for Excellent Young Scholars of Northeast Institute of Geography and Agroecology, CAS (DLSYQ13004).

Acknowledgments: The authors also want to provide our gratitude to the editors and the anonymous reviewers who gave us their insightful comments and suggestions.

Conflicts of Interest: The authors declare no conflict of interest.

References

1. Cao, S.; Chen, L.; Liu, Z. An investigation of Chinese attitudes towards the environment: Case study using the Grain for Green Project. *Ambio* **2009**, *38*, 55–64. [CrossRef] [PubMed]
2. McPherson, E.G.; Simpson, J.R.; Peper, P.J.; Maco, S.E.; Xiao, Q. Municipal forest benefits and costs in five U.S. cities. *J. For.* **2005**, *103*, 411–416.
3. Nowak, D.J.; Greenfield, E.J.; Hoehn, R.E. Carbon storage and sequestration by trees in urban and community areas of the United States. *Environ. Pollut.* **2013**, *178*, 229–236. [CrossRef] [PubMed]
4. Puneet, D.; Chinmaya, S.R.; Yogesh, D. Ecological benefits of urban forestry: The case of Kerwa Forest Area (KFA), Bhopal, India. *Appl. Geogr.* **2009**, *29*, 194–200.
5. Young, R.F. Managing municipal green space for ecosystem services. *Urban For. Urban Green.* **2010**, *9*, 313–321. [CrossRef]
6. Nowak, D.J.; Crane, D.E. Carbon storage and sequestration by urban trees in the USA. *Environ. Pollut.* **2002**, *116*, 381–389. [CrossRef]
7. Weng, Q.; Yang, S. Urban air pollution patterns, land use, and thermal landscape: An examination of the linkage using GIS. *Environ. Monit. Assess.* **2006**, *117*, 463–489. [CrossRef] [PubMed]
8. Hall, R.J.; Skakun, R.S.; Arsenault, E.J.; Case, B.S. Modeling forest stand structure attributes using Landsat ETM+ data: Application to mapping of aboveground biomass and stand volume. *For. Ecol. Manag.* **2006**, *225*, 378–390. [CrossRef]
9. Lucy, R.H.; Byungman, Y.; Marina, A. Terrestrial carbon stocks across a gradient of urbanization: A study of the Seattle, WA region. *Glob. Chang. Biol.* **2010**, *17*, 783–797.
10. Kirnbauer, M.C.; Baetz, B.W.; Kenney, W.A. Estimating the stormwater attenuation benefits derived from planting four monoculture species of deciduous trees on vacant and underutilized urban land parcels. *Urban For. Urban Green.* **2013**, *12*, 401–407. [CrossRef]
11. Armson, D.; Stringer, P.; Ennos, A.R. The effect of street trees and amenity grass on urban surface water runoff in Manchester, UK. *Urban For. Urban Green.* **2004**, *12*, 282–286. [CrossRef]
12. Bowler, D.E.; Buyung-Ali, L.; Knight, T.M.; Pullin, A.S. Urban greening to cool towns and cities: A systematic review of the empirical evidence. *Landsc. Urban Plan.* **2010**, *97*, 147–155. [CrossRef]
13. Shashua-Bar, L.; Hoffman, M.E. Vegetation as climatic component in the design of an urban street—An empirical model for predicting the cooling effect of urban green areas with trees. *Energy Build.* **2000**, *31*, 221–235. [CrossRef]
14. Cornelis, J.; Hermy, M. Biodiversity relationships in urban and suburban parks in Flanders. *Landsc. Urban Plan.* **2004**, *69*, 385–401. [CrossRef]
15. Godefroid, S.; Koedam, N. How important are large vs. small forest remnants for the conservation of the woodland flora in an urban context? *Glob. Ecol. Biogeogr.* **2003**, *12*, 287–298. [CrossRef]
16. Mcpherson, E.G.; Nowak, D.; Heisler, G. Quantifying urban forest structure, function, and value: The Chicago Urban Forest Climate Project. *Urban Ecosyst.* **1997**, *1*, 49–61. [CrossRef]
17. Nowak, D.J. Understanding the structure of urban forests. *J. For.* **1994**, *92*, 36–41.

18. Wu, J. The state-of-the-science in urban ecology and sustainability. *Landsc. Urban Plan.* **2014**, *125*, 209–221. [CrossRef]
19. Tian, G.; Liu, J.; Xie, Y.; Yang, Z.; Zhuang, D.; Niu, Z. Analysis of spatio-temporal dynamic pattern and driving forces of urban land in China in 1990s using TM images and GIS. *Cities* **2005**, *22*, 400–410. [CrossRef]
20. Zhou, X.; Wang, Y. Spatial–temporal dynamics of urban greenspace in response to rapid urbanization and greening policies. *Landsc. Urban Plan.* **2011**, *100*, 268–277. [CrossRef]
21. Yang, J.; Huang, C.; Zhang, Z.; Wang, L. The temporal trend of urban green coverage in major Chinese cities between 1990 and 2010. *Urban For. Urban Green.* **2013**, *13*, 19–27. [CrossRef]
22. Zhao, J.; Chen, S.; Jiang, B.; Ren, Y.; Wang, H.; Vause, J.; Yu, H. Temporal trend of green space coverage in China and its relationship with urbanization over the last two decades. *Sci Total Environ.* **2013**, *442*, 455–465. [CrossRef] [PubMed]
23. Liu, C.F.; Li, X.M. Carbon storage and sequestration by urban forests in Shenyang, China. *Urban For. Urban Green.* **2012**, *11*, 121–128. [CrossRef]
24. Zhang, D.; Zheng, H.; Ren, Z.; Zhai, C.; Shen, G.; Mao, Z.; Wang, P.; He, X. Effects of Forest Type and Urbanization on Carbon Storage of Urban Forests in Changchun, Northeast China. *Chin. Geogr. Sci.* **2015**, *25*, 147–158. [CrossRef]
25. Escobedo, F.; Varela, S.; Zhao, M.; Wagner, J.E.; Zipperer, W. Analyzing the efficacy of subtropical urban forests in offsetting carbon emissions from cities. *Environ. Sci. Policy* **2010**, *13*, 362–372. [CrossRef]
26. Frolking, S.; Palace, M.; Clark, D.B.; Chambers, J.Q.; Shugart, H.H.; Hurtt, G.C. Forest disturbance and recovery—A general review in the context of space-borne remote sensing of impacts on aboveground biomass and canopy structure. *J. Geophys. Res.* **2009**, *114*, 281–296. [CrossRef]
27. Ingram, J.C.; Terence, P.; Dawson, R.J. Mapping tropical forest structure in southeastern Madagascar using remote sensing and artificial neural networks. *Remote Sens. Environ.* **2005**, *94*, 491–507. [CrossRef]
28. Simone, R.F.; Marcia, C.; Carla, B. Relationships between forest structure and vegetation indices in Atlantic Rainforest. *For. Ecol. Manag.* **2005**, *218*, 353–362.
29. Ji, L.; Bruce, K.W.; Dana, R.N. Estimating aboveground biomass in interior Alaska with Landsat data and field measurements. *Int. J. Appl. Earth Obs.* **2012**, *18*, 451–461. [CrossRef]
30. Lu, D.S.; Paul, M.; Eduardo, B.; Emilio, M. Relationships between forest stand parameters and Landsat TM spectral responses in the Brazilian Amazon Basin. *For. Ecol. Manag.* **2004**, *198*, 149–167. [CrossRef]
31. Huang, X.; Huang, X.J.; Chen, C. The Characteristic, Mechanism and Regulation of Urban Spatial Expansion of Changchun. *Areal Res. Dev.* **2009**, *5*, 68–72.
32. Chander, G.; Markham, B. Revised Landsat-5 TM radiometric calibration procedures and postcalibration dynamic Ranges. *IEEE Trans. Geosci. Remote Sens.* **2003**, *41*, 2674–2677. [CrossRef]
33. Schott, J.R.; Salvaggio, C.; Volchok, W.J. Radiometric scene normalization using pseudoinvariant features. *Remote Sens. Environ.* **1988**, *26*, 1–16. [CrossRef]
34. Yang, X.; Lo, C.P. Relative radiometric normalization performance for change detection from multi-date satellite images. *Photogramm. Eng. Remote Sens.* **2000**, *66*, 967–980.
35. Nowak, D.J.; Crane, D.E.; Stevens, J.C.; Hoehn, R.E. *The Urban Forest Effects (UFORE) Model: Field Data Collection Manual*; US Department of Agriculture Forest Service, Northeastern Research Station: Syracuse, NY, USA, 2003.
36. Liu, C.F.; Li, M.M.; He, X.Y.; Chen, W.; Xu, W.Y.; Zhao, G.L.; Ning, Z.H. Concept discussion and analysis of urban forest. *Chin. J. Ecol.* **2003**, *22*, 146–149. (In Chinese)
37. Zhou, W.; Troy, A. Development of an object-based framework for classifying and inventorying human-dominated forest ecosystems. *Int. J. Remote Sens.* **2009**, *30*, 6343–6360. [CrossRef]
38. Gao, Y.; Mas, J.F. A comparison of the performance of pixel based and object based classifications over images with various spatial resolutions. *J. Earth Sci.* **2008**, *2*, 27–35.
39. Baret, F.; Guyot, G. Potentials and limits of vegetation indexes for LAI and APAR assessment. *Remote Sens. Environ.* **1991**, *35*, 161–173. [CrossRef]
40. Gower, S.T.; Kucharik, C.J.; Norman, J.M. Direct and indirect estimation of leaf area index, fAPAR, and net primary production of terrestrial ecosystems. *Remote Sens. Environ.* **1999**, *70*, 29–51. [CrossRef]
41. Gray, J.; Song, C.H. Mapping leaf area index using spatial, spectral, and temporal information from multiple sensors. *Remote Sens. Environ.* **2012**, *119*, 173–183. [CrossRef]

42. Franklin, J.; Hiernaux, P.H. Estimating foliage and woody biomass in Sahelian and Sudanian woodlands using a remote sensing model. *Int. J. Remote Sens.* **1991**, *12*, 1387–1404. [CrossRef]
43. Dallimer, M.; Tang, Z.; Bibby, P.R.; Brindley, P.; Gaston, K.J.; Davies, Z.G. Temporal changes in greenspace in a highly urbanized region. *Biol. Lett.* **2011**, *7*, 763–766. [CrossRef] [PubMed]
44. Baycan-Levent, T.; Vreeker, R.; Nijkamp, P.A. The multi-criteria evaluation of greenspaces in European cities. *Eur. Urban Reg. Stud.* **2009**, *16*, 193–213. [CrossRef]
45. Kabisch, N.; Haase, D. Green spaces of European cities revisited for 1990–2006. *Landsc. Urban Plan.* **2013**, *110*, 113–122. [CrossRef]
46. Nowak, D.J.; Greenfield, E.J. Tree and impervious cover in the United States. *Landsc. Urban Plan.* **2012**, *107*, 21–30. [CrossRef]
47. Chinese Urban Forest Website. Available online: http://www.cuf.com.cn/ (accessed on 12 November 2017).
48. Hubacek, K.; Guan, D.; Barrett, J.; Wiedmann, T. Environmental implications of urbanization and lifestyle change in China:Ecological and Water Footprints. *J. Clean. Prod.* **2009**, *17*, 1241–1248. [CrossRef]
49. Gómez-Baggethun, E.; Barton, D.N. Classifying and valuing ecosystem services for urban planning. *Ecol. Econ.* **2013**, *86*, 235–245. [CrossRef]
50. Deal, B.; Pan, H. Discerning and Addressing Environmental Failures in Policy Scenarios Using Planning Support System (PSS) Technologies. *Sustainability* **2017**, *9*, 13. [CrossRef]
51. .Landry, S.M.; Chakraborty, J. Street trees and equity: Evaluation the spatial distribution of an urban amenity. *Environ. Plan.* **2009**, *41*, 2651–2670. [CrossRef]
52. Tooke, T.R.; Klinkenberg, B.; Coops, N.C. A geographical approach to identifying vegetation-related environmental equity in Canadian cities. *Environ. Plan.* **2010**, *37*, 1040–1056. [CrossRef]

© 2018 by the authors. Licensee MDPI, Basel, Switzerland. This article is an open access article distributed under the terms and conditions of the Creative Commons Attribution (CC BY) license (http://creativecommons.org/licenses/by/4.0/).

Article

Urban Park Systems to Support Sustainability: The Role of Urban Park Systems in Hot Arid Urban Climates

Gunwoo Kim [1],* and Paul Coseo [2]

1. Landscape Architecture Program, North Carolina A&T State University, 231 D Carver Hall, 1601 E. Market Street, Greensboro, NC 27411, USA
2. Landscape Architecture Program, Arizona State University, DN 394, 810 S. Forest Ave., P.O. Box 871605, Tempe, AZ 85287-1605, USA; Paul.Coseo@asu.edu
* Correspondence: gkim1@ncat.edu

Received: 18 June 2018; Accepted: 20 July 2018; Published: 22 July 2018

Abstract: Quantifying ecosystem services in urban areas is complex. However, existing ecosystem service typologies and ecosystem modeling can provide a means towards understanding some key biophysical links between urban forests and ecosystem services. This project addresses broader concepts of sustainability by assessing the urban park system in Phoenix, Arizona's hot urban climate. This project aims to quantify and demonstrate the multiple ecosystem services provided by Phoenix's green infrastructure (i.e., urban park system), including its air pollution removal values, carbon sequestration and storage, avoided runoff, structural value, and the energy savings it provides for city residents. Modeling of ecosystem services of the urban park system revealed around 517,000 trees within the system, representing a 7.20% tree cover. These trees remove about 3630 tons (t) of carbon (at an associated value of $285,000) and about 272 t of air pollutants (at an associated value of $1.16 million) every year. Trees within Phoenix's urban park system are estimated to reduce annual residential energy costs by $106,000 and their structural value is estimated at $692 million. The findings of this research will increase our knowledge of the value of green infrastructure services provided by different types of urban vegetation and assist in the future design, planning and management of green infrastructure in cities. Thus, this study has implications for both policy and practice, contributing to a better understanding of the multiple benefits of green infrastructure and improving the design of green spaces in hot arid urban climates around the globe.

Keywords: ecosystem services; ecosystem modeling; sustainability; human health; environmental quality; hot arid urban climate

1. Introduction

Urban forests within urban park systems provide essential buffering ecosystem services for the sustainability of rapidly urbanizing cities. Ecosystem services are those ecological functions, provided by ecosystems for free, that society values for their supply of public goods including ecosystems' ability to provide essential provisioning, regulating, supporting, and cultural services [1,2]. Rapid urbanization, in places like Phoenix, Arizona (AZ), delete or degrade endemic ecosystems and replace them with novel biogeophysical compositions and configurations. Rapid urbanization is a global phenomenon that also results in more people living in cities being exposed to hazards from urbanization. From 1950 to 2010, urban residents increased from 29% to 50% globally and by 2050 the proportion of urban residents is expected to reach 69% of the global population, which represents 6.3 billion people. Thus, urban forests are key to protecting the health and well-being for most people around the world [3]. Urban parks are dedicated islands within urbanized areas to nurture much-needed ecosystem services in cities. Other urbanized lands are dedicated to grey infrastructure (i.e., buildings and pavement) that are primarily composed of mineral-based materials, which seal soils and alter important biogeochemical processes. City residents need these critical urban park systems to cool and clean the air, sequester and store carbon, absorb and clean runoff, enhance biodiversity, and reduce energy demands [4]; all problems resulting from grey infrastructure. For example, urban-induced warming (i.e., urban heat island effect) results from constructing buildings and pavements, replacing unbuilt landscapes that previously maintained endemic heating and cooling processes [5].

In rapidly urbanizing cities, like the Phoenix Metropolitan Area (PMA), urbanization has already caused some impoverished, tree-poor, compact neighborhoods' temperatures to be up to 6.4 °C warmer than better-off neighborhoods with a wealth of trees and vegetation [6]. The pace of urbanization is intimately linked with urban-induced warming and has been found to accelerate the urban heat island (UHI) phenomena [5]. Brazel and colleagues [7] found that in Phoenix, rapid urbanization, warmed the air temperature in June by 1.4 °C per 1000 homes constructed. Urban-induced warming is made worse when combined with global climate change due to increasing concentrations of greenhouse gases. These two mechanisms will continue to warm rapidly growing cities during this century. On top of urban-induced warming, the continental U.S. average temperatures have increased by 1.3 °F to 1.9 °F since record keeping began in 1985. This global warming has resulted in more intense and frequent heat waves, and the future likely holds more extreme heat. Annual average U.S. temperatures are projected to increase by 3 °F to 10 °F by the end of the 21st century, depending on future emissions of greenhouse gases and other regional factors [8]. Together, warming from both urban-induced and global temperature increases has the potential to make many urban neighborhoods unsustainable. People residing in hot neighborhoods are more vulnerable to heat exposure because they have fewer social and material resources to cope with extreme heat [6]. Cities around the world are facing various urban problems driven by increasing urban populations, urban-induced warming, and climate change [9]; urban park systems play a key role in addressing those negative urban problems.

Urbanization is key driver of changes in ecosystem services. For these reasons, policy makers, academics, urban planners and engineers have started to focus on the ecosystem services required to move toward urban sustainability [10]. Yet, advocates require more convincing and empirically driven evidence on the social, environmental, and economic return on capital investments in propagating and managing healthy urban forests. In arid cities, it is not just an investment in urban forests, but also the water needed to maintain those forests. Stronger justification and evidence are needed in arid cities where water availability is a critical sustainability issue. Urban forests in cities across the country are important for sustainable futures by improving numerous environmental and social aspects of cities, including human health, walkability, thermal comfort, cultural desires, stormwater management, air quality, wildlife habitat, aesthetics and carbon sequestration. For urban forestry, the challenge cities face includes understanding and clearly illustrating the value trees provide to society and how cities should fund effective urban forestry management programs. Quantifying ecosystem services in urban areas is complex. However, existing ecosystem service typologies and ecosystem modeling can provide a means towards understanding some key biogeophysical links between urban forest and ecosystem services [11]. The overall objective of this project is to improve our understanding of the complex interrelationships between ecosystem services, human health and well-being to craft multiple forestry strategies that support sustainable futures. This project addresses the broader concept of sustainability by assessing the value of an urban park system in Phoenix, Arizona (AZ), USA.

The purpose of this study is to quantify and demonstrate the multiple ecosystem services provided by Phoenix's green infrastructure (i.e., urban park system), including air pollution removal, carbon sequestration and storage, avoided runoff, structural value, and energy savings for residents in the city's hot arid urban climate. In particular, Phoenix's hot arid urban conditions provide a glimpse at what many rapid urbanizing cities around the globe may face from global climate change and urban-induced warming in the 21st Century. This study will assess the status of Phoenix's existing urban park system's forest to estimate the environmental benefits and ecosystem services provided, thus improving our understanding of the role trees play in creating healthy, livable, and sustainable cities [12]. Modelling urban forest structure provides useful information for estimating the total leaf area, tree and leaf biomass, and quantifying the numerous ecosystem services and forest functions; accurate assessments are critical for managers and planners to understand how the various ecosystem services improve both environmental quality and human health and well-being in urban areas [13]. Though urban forests perform many functions and add value to many aspects of everyday life, currently only a few of these attributes can be accessed due to our limited ability to quantify all of these values utilizing standard data analyses [14].

The most precise way to assess urban forests is to measure and record every tree on a site, but although this works well for relatively small areas such as a single street with trees and small parks it is prohibitively expensive for large tree populations [13]. Instead, random sampling techniques offer a cost-effective way to assess urban forest structure and ecosystem services for large-scale assessments [13]. Comprehensive assessments of urban forest structure are thus commonly conducted using sampling techniques (e.g., [15–20]). To facilitate this approach, the U.S. Forest Service's Northern Research Station developed the i-Tree Eco model (formerly known as the Urban Forest Effects model) to support assessments of urban forest structure, ecosystem services, and economic benefits [21]. The i-Tree Eco model incorporates protocols to measure and monitor urban forest structure and estimate its ecosystem functions and economic values; the associated software utilizes standardized field data from randomly located plots and local hourly air pollution and meteorological data to quantify urban forest structure and its numerous effects [13]. The i-Tree model is already used in over 50 cities to assess urban forest structure and functions using a standardized approach (see, for example [20,22,23]). The study utilized i-Tree Eco to assess the urban park system's forest structure and ecosystem services in the City of Phoenix, AZ, USA.

2. Methods

In terms of the science and mechanics of modeling ecosystem services, the starting point is understanding forest structure—the extent, distribution and composition of urban forest. Urban forest assessment is essential for developing a baseline from which to measure changes and trends. Managing the urban forest includes tree maintenance, policy development, and budgetary decisions—all of which depend on understanding current urban forest conditions [12]. There are two ways of assessing urban forest structure using both on-the-ground measurements and remote sensing analysis. An accurate quantification of urban forests can help in understanding the various ecosystem services and values it provides [13]. Top-down (aerial) approaches produce good cover estimates and can detail and map tree and other cover locations; however, they tell little about composition and therefore little about services provided. Bottom-up (field inventory) approaches provide detailed management information, such as number of trees, species composition, tree sizes and health, tree locations, and risk information. This approach provides better means to assess and project ecosystem services and values into the future. This bottom-up approach is the foundation of i-Tree Eco and thus was the approach utilized for this study.

i-Tree Eco allows users to inventory or sample tree populations anywhere and at any scale (single tree to large region) to estimate tree populations (e.g., number of trees, species composition). It then combines local weather and pollution data to estimate ecosystem services and values by simulating the trees under local conditions using local tree data. The i-Tree Eco software is designed to apply standardized field data from randomly located plots and local hourly air pollution measures and meteorological data (e.g., volatile organic compound emission, air pollution removal by the urban forest, relative ranking of species' effects on air quality, tree transpiration) to quantify: (1) the structure of urban forests and the resulting effects on local air quality; (2) runoff, stream flow and water quality; (3) building energy use; (4) carbon sequestration; and (5) air temperatures. The methodological framework of the i-Tree Eco model is based on an assessment of the urban forest structure, function, and value.

In the City of Phoenix, 50 (0.04 ha) plots were sampled using a random sampling method across the urban park system made up of both publicly- and privately-owned land parcels (Figure 1 and Table 1). Phoenix has a number of master plan gated communities that represent the privately owned and managed neighborhood parks that were included in the study. Plots were permanently referenced so that they can be monitored over time and were assigned proportionate to the land area within each stratum based on land use zoning. All field data were collected during the 2016 vegetation period (May–June) to properly assess the tree canopy health. At each field plot, three crew members collected data on ground vegetation and tree cover, individual tree attributes such as species, stem diameter, height, crown width, crown canopy missing and dieback, and their distance and orientation to neighboring buildings [13]. Field data was input into the i-Tree Eco model to assess forest structure and the associated ecosystem services and values. The initial field data analysis was conducted using i-Tree Eco with assistance from scientists and staff at the United States Department of Agriculture (USDA) Forest Service's Northern Research Station and the Davey Tree Expert Company. Additional analyses were undertaken by project researchers to explore information of specific interest for Phoenix. Details of the i-Tree Eco methods are available at the i-Tree website (www.itreetools.org) and in several publications (e.g., [13,24,25]).

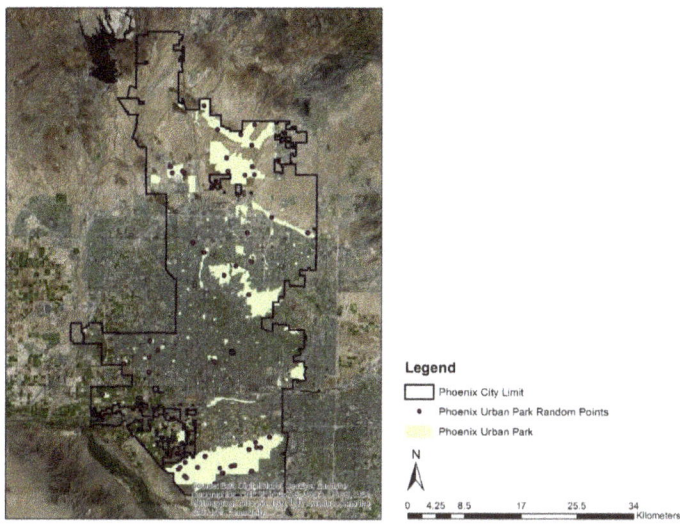

Figure 1. Urban park system plot locations within publicly and privately owned land parcels, City of Phoenix, Arizona, USA.

Hourly weather data are collected to analyze air pollution removal by the urban forest, such as volatile organic compound emissions, air pollution removal by the urban forest, relative ranking of species' effects on air quality and tree transpiration [24]. Air pollution removal estimates are derived from calculated hourly tree-canopy resistance for ozone, and sulfur and nitrogen dioxides based on a hybrid of the big-leaf and multi-layer canopy deposition models [26,27]. Recent updates to the air quality models are based on improved leaf area index simulations, weather and pollution processing and interpolation, and updated pollutant monetary values. To calculate current carbon storage levels, the biomass for each tree is calculated using published allometric equations and measures of tree data [22,28]. Carbon storage and carbon sequestration values are based on estimated or customized local carbon values. Carbon storage and carbon sequestration values in Phoenix are calculated based on $78 per ton [29].

Annual avoided surface runoff is calculated based on rainfall interception by vegetation, specifically the difference between annual runoff with and without vegetation. Estimates are made of the percent of the area beneath the dripline of the tree that is impervious or occupied by shrubs. Although tree leaves, branches, and bark may intercept precipitation and thus mitigate surface runoff, only the precipitation intercepted by leaves will be taken into account in this study. The value of avoided runoff is based on estimated local values based on the United States Forest Service's Community Tree Guide Series [30]. Structural values were based on the valuation procedures utilized by the Council of Tree and Landscape Appraisers, which uses tree species, diameter, condition, and location information [31]. Structural value is based on four tree/site characteristics: (1) trunk area (cross-sectional area at diameter at breast height); (2) species; (3) condition; and (4) location. Trunk area and species will be used to determine the basic value, which is then multiplied by condition and location ratings (0 to 1) to determine the final tree compensatory value. The seasonal effects of trees on residential building energy use are calculated based on procedures described in the literature [32] using the distance and direction of trees from residential structures, tree height, and tree condition data. To calculate the monetary value of energy savings achieved, local or custom prices per Megawatt-hour (MWH) or One million British Thermal Units (MBTU) are utilized. One measure of the relative dominance of species in a forest community is called the Importance Value (IV). IV ranks species within a site based upon three criteria: (1) how commonly a species occurs across the entire forest; (2) the total number of individuals of the species; and (3) the total amount of forest area occupied by the species [33]. IV is calculated as the sum of relative leaf area and relative composition. To compare forest communities' composition that may differ in size, or that were sampled at different intensities, IVs are calculated using relative rather than absolute values.

Urban forest ecosystem functions' quantification procedures are estimated based on various algorithms and many of the ecosystem functions estimated by the i-Tree Eco model are difficult to accurately measure in the field; thus, modeling procedures are needed to quantify these effects for urban forests [13]. Due to the importance of the quality assurance of field data accuracy, the model estimates are only as good as the field data inputs; the i-Tree Eco model estimates current urban forest structure and functions and then treats this as a permanent average value for the plot [13]. Urban forest conditions are changeable, so the model value is not absolute; it represents a snap shot in time. The precision and cost of the estimate is also dependent on the sample and plot size. Generally, 200 plots (0.04 ha each) in a stratified random sample (with at least 15 plots per stratum) will produce a 12% relative standard error for an estimate covering the entire study area [25]. As the number of plots increases, the standard error decreases and the method provides more accurate population estimates. However, as the number of plots increases, so does the time and cost of field data collection.

Table 1. Current urban land area and percentage with completed plots within the urban park system of Phoenix, Arizona, USA (total area: 1341.48 km^2).

Typology of Green Infrastructure	Existing Green Infrastructure Land			Number of Plots Selected for Analysis
	Area (km^2)	±SE	% of Total Area	
Urban park system	194.45	±5.06	14.5%	50
City Total	1341.48		100%	

±SE = Standard error of the total.

3. Results

3.1. Urban Forest Structure of Phoenix's Park System

The forest structure of Phoenix's park system is the amount and density of plants, the types of plants present (e.g., trees, shrubs, ground cover), the diversity of species, and tree health. The urban forest of Phoenix's park system has an estimated 517,000 trees with a tree cover of 7.20%. Figure 2 illustrates the most common tree species growing in the park system. The top three species are *Vachellia farnesiana* (Sweet Acacia) (25.6%), *Parkinsonia microphylla* (Yellow Paloverde) (16.3%), and *Prosopis velutina* (Velvet Mesquite) (14.0%). The majority of trees growing within the park system had diameters less than 15.2 cm constituting 53.3% of the tree population (Figure 3). The overall tree density of the park system was 26.6 trees per ha (Table 2).

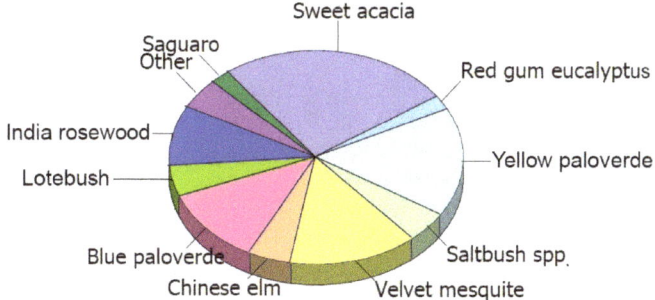

Figure 2. Tree species composition in Phoenix's urban park system.

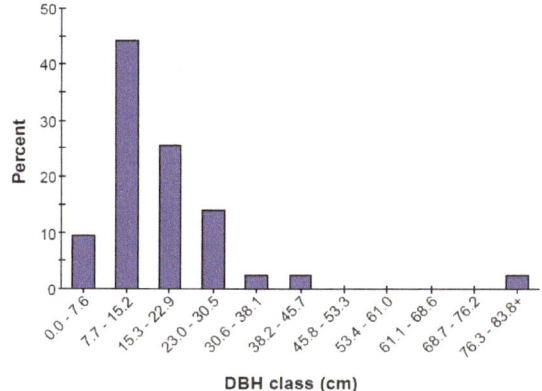

Figure 3. Percentage of tree population by diameter (DBH = stem diameter at 1.37 m above the ground line).

Table 2. Percentage tree cover and number of trees within Phoenix's urban park system.

Typology of Green Infrastructure	Area (km²) (SE)	Percentage Tree Cover (SE)	Number of Trees (SE)	Number of Trees per ha (SE)
Urban park system	194.45	7.2 (2.68)	516,534 (197,947)	26.6 (10.18)

SE = Standard error of total.

The urban forest is composed of a mix of native and exotic tree species. Thus, Phoenix's park system has a tree diversity that is higher than surrounding native landscapes. In Phoenix's park system, about 65% of the trees are species native to North America, while 56% are native to the state or the Lower Colorado River Valley ecosystem. Species exotic to North America make up 35% of the tree population. Most exotic tree species have an origin from North & South America (14% of the species) (Figure 4). Invasive plant species are often characterized by their vigor, ability to adapt, reproductive capacity, and general lack of natural enemies. These abilities enable them to displace native plants and make them a threat to natural areas [34]. None of the 12 tree species sampled in Phoenix's park system is defined as invasive on the state invasive species list [35] (Table 3).

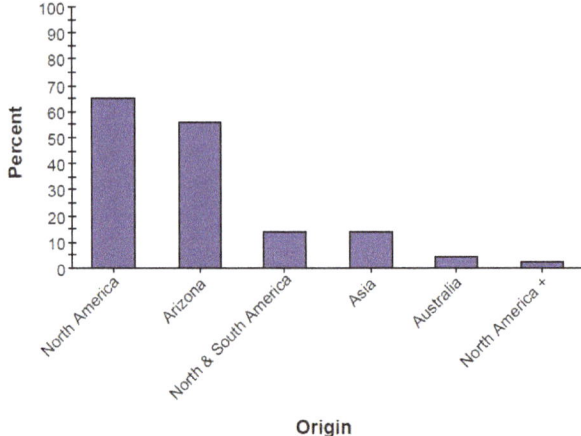

Figure 4. Percentage of live tree species by origin area growing in Phoenix's park system. The plus sign (+) indicates the plant is native to another continent other than the ones listed in the grouping.

Table 3. Urban forest in Phoenix: tree biodiversity within the urban park system.

Typology of Green Infrastructure	Number of Tree Species	Number of Native Species	Number of Non-Native Species	Number of Invasive Species
Urban park system	12	8	4	0

3.2. Urban Forest Cover and Leaf Area on Phoenix's Park System

Ecosystem services are directly proportional to the total number of trees, species, percentage of tree canopy cover and healthy leaf surface area of the plant. The amount of forest cover is a critical cooling mechanism for both its shade and its evapotranspiration. In Phoenix, the three most dominant tree species in terms of leaf area are *Vachellia farnesiana* (Sweet Acacia), *Eucalyptus camaldulensis* (Red Gum Eucalyptus), and *Ulmus parvifolia* (Chinese Elm) (Table 4). Tree canopy covers about 7.2% of Phoenix's park system. Importance values (IV) are calculated as the sum of relative leaf area and relative composition. An IV over 10 may indicate that the park system is over-reliant on a particular species for structural and functional benefits, depending on the local ecosystem [36]. Phoenix's park system has four species with an IV exceeding 10, the most important of which is Sweet Acacia with an IV of 48.5. The 10 most important species within Phoenix's park system are listed in Table 4.

Table 4. Most important species within the urban park system, Phoenix, Arizona.

Species Name	Percentage of Population	Percentage of Leaf Area	Importance Value (IV)
Sweet acacia	25.6	22.9	48.5
Yellow paloverde	16.3	12.7	29.0
Red gum eucalyptus	2.3	21.1	23.4
India rosewood	9.3	13.9	23.2
Chinese elm	4.7	14.0	18.7
Velvet mesquite	14.0	3.8	17.8
Blue paloverde	11.6	3.9	15.5
Saltbush spp.	4.7	2.4	7.1
Bottle tree	2.3	3.5	5.8
Lotebush	4.7	0.9	5.6

Estimating land cover types of the urban park system is another important element to assess ecosystem services of those parks. Bare soil (43.2%) and turf grass (26.6%) are the two most dominant permeable land cover types in Phoenix's park system (Table 5), which means that Phoenix can use its park system to strategically control and manage urban stormwater. The two impervious land cover classes (concrete/asphalt pavement and rock) make up a small percentage (15.4%) of the urban park system's total land area (Table 5). The plantable space (not covered by impervious surfaces and free of existing tree canopy cover) represents a high percentage (64.1%) of the park system, which suggests that parks have a high potential for increasing Phoenix's tree canopy cover. Phoenix has ample room to grow and expand its urban forest through its park system. As the tree canopy cover increases, this also provides other ecosystem services and benefits to local residents.

Table 5. Percentage of land cover in the urban park system, City of Phoenix, Arizona.

Typology of Green Infrastructure	Plant Space	Land Cover								
		Concrete/Asphalt Pavement	Tar	Bare Soil	Rock	Herbs	Grass	Wild Grass	Water	Shrub	Tree
Urban park system	64.1	9.5	1.5	43.2	5.9	2.4	26.6	4.0	0.1	6.8	7.2

Land cover totals 100% and includes pavement, tar, bare soil, rock, herbs, grass, wild grass, water, and shrub. Plant space and tree cover overlap with land cover.

3.3. Air Pollution Removal by Phoenix's Park System

Trees within urban park systems can help improve air quality by directly removing pollutants from the air, reducing ambient air temperature through shade and transpiration, and reducing energy consumption in buildings through shade. This reduced energy consumption also reduces waste heat and air pollutant emissions from air conditioning units and power plants. Recent updates to the air quality models are based on improved leaf area index simulations, weather and pollution processing and interpolation, and updated pollutant monetary values. As shown in Figure 5, ozone (O_3) had the greatest pollution removal value. Overall, 272 t of air pollutants (CO, NO_2, O_3, PM_{10}, and SO_2) were removed by trees within Phoenix's park system with an annual value of $1.16 million. The urban park system is a critical component of Phoenix's green infrastructure to enhance ecosystem and human health, thus promoting better quality of life for city residents.

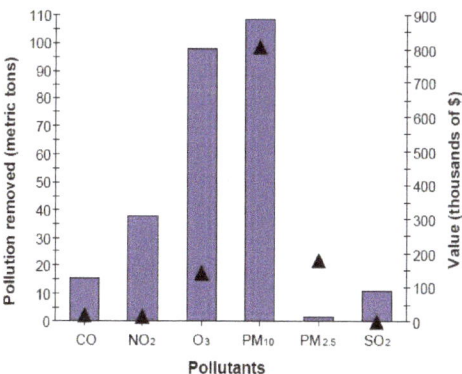

Figure 5. Pollution removal (bars) and associated value (points) for trees in Phoenix's park system. Pollution removal values were calculated based on the prices of $1253 per metric ton (CO, carbon monoxide), $1472 per metric ton ($O_3$, ozone), $355 per metric ton ($NO_2$, nitrogen dioxide), $93 per metric ton ($SO_2$, sulfur dioxide), $7425 per metric ton ($PM_{10}$, particulate matter less than 10 microns and greater than 2.5 microns), $105,201 per metric ton ($PM_{2.5}$, particulate matter less than 2.5 microns).

3.4. Carbon Storage and Sequestration on Phoenix's Park System

Urban forests within urban park systems can help mitigate climate change by sequestering atmospheric carbon (from carbon dioxide) in tissue and by reducing energy use in buildings, and consequently lowering carbon dioxide emissions from fossil-fuel based power [37]. The gross sequestration of trees within Phoenix's park system is about 3630 t of carbon per year (Table 6), with an associated value of $285,000. Net carbon sequestration (accounting for losses from carbon dioxide release through tree respiration) in green infrastructure is about 3380 t annually (Table 6). Trees store and sequester carbon dioxide through growth processes in their tissue. Carbon storage and carbon sequestration values are based on estimated or customized local carbon values. Carbon storage and carbon sequestration values in Phoenix are calculated based on $78 per metric ton [29].

Table 6. City total for tree effects by Phoenix's park system.

Typology of Green Infrastructure	Percentage Tree Cover (SE)	Number of Trees (SE)	Accumulated Carbon Storage (t) (SE)	Gross Carbon Sequestration (t/year) (SE)	Net Carbon Sequestration (t/year) (SE)
Urban park system	7.2 (2.68)	516,534 (197,947)	57,755.8 (31,396.3)	3633.3 (1288.4)	3383.5 (1171.1)

SE = Standard error of total.

Urban forests supported by Phoenix's park system are estimated to store 57,800 t of carbon, which is valued at $4.53 million (Figure 6). Of all the species sampled, *Eucalyptus camaldulensis* Dehnh. (Red Gum Eucalyptus) stores and sequesters the most carbon (approximately 51.9% of the total carbon stored and 24.7% of all carbon sequestered in trees growing in parks) (Figure 6).

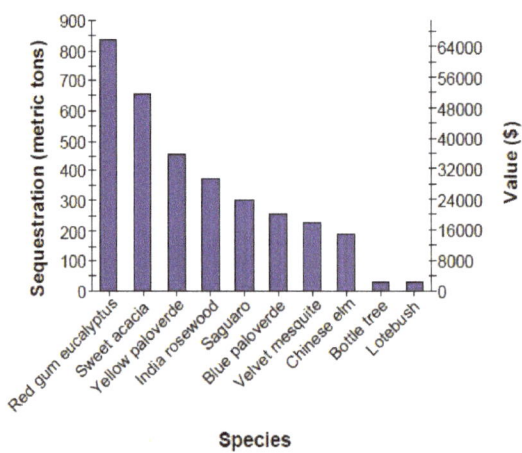

Figure 6. Carbon sequestration and value for the species with greatest overall carbon sequestration growing within Phoenix's park system.

3.5. Avoided Runoff on Phoenix's Park System

In Phoenix, the urban park system has two impervious ground cover classes (pavement and rock), which make up 15.4% of the total land cover in this category (Table 5) and has 7.2% of the tree canopy cover. The plantable space available on the urban park system is about 64.1%, which is high relative to other land covers [36]. Therefore, the urban park system has considerable additional capacity to reduce surface runoff if strategies are implemented within the system to capture stormwater in tree bioretention and infiltration beds. This suggests that Phoenix's park system could be a more valuable ecological resource that can be strategically used to increase sustainability to extreme rainfall events through urban green stormwater infrastructure that simultaneously nurtures urban forests, including trees, shrubs and pervious land cover classes. For example, urban trees within park systems are highly beneficial in reducing surface runoff. Trees intercept precipitation, while their root systems promote infiltration and storage in the soil. Currently, it's estimated that the trees growing in the City's park system help to reduce runoff by an estimated 52,800 cubic meters a year, with an associated value of $124,000, as shown in Table 7 and Figure 7 [30].

Table 7. City total for avoided runoff by Phoenix's park system.

Typology of Green Infrastructure	Number of Trees (SE)	Leaf Area Area (km²) (SE)	Avoided Runoff (m³/year)	Avoided Runoff Value ($)
Urban park system	516,534 (197,947)	45.8 (15.7)	52,791.65	124,120.11

SE = Standard error of total; Avoided runoff is calculated by the price $2.351/m³.

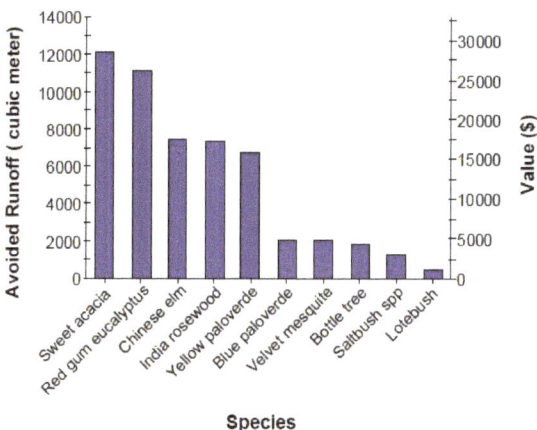

Figure 7. Avoided runoff and values for species with the greatest overall impact on runoff within Phoenix's park system.

3.6. Phoenix's Park System and Building Energy Use

Based on state-wide energy costs for Phoenix ($115.3 per MWH and $16.8 per MBTU), the trees growing within its park system were estimated to reduce energy-related costs from residential buildings by $106,000 annually. Trees also provided an additional $12,482 in value by reducing the amount of carbon released by fossil-fuel based power plants (a reduction of 159 t of carbon emissions) (Tables 8 and 9).

Table 8. Annual energy conservation and carbon avoidance due to trees within Phoenix's park system near residential buildings (note: negative numbers indicate an increased energy use or carbon emission).

	Heating	Cooling	Total
MBTU [1]	350	n/a	350
MWH [2]	29	839	868
Carbon avoided (mt [3])	8	151	159

[1] One million British Thermal Units; [2] Megawatt-hour; [3] Metric ton.

Table 9. Annual savings [1] ($) in residential energy expenditure during heating and cooling seasons (note: negative numbers indicate a cost due to increased energy use or carbon emission).

	Heating	Cooling	Total
MBTU [2]	5880	n/a	5880
MWH [3]	3344	96,737	100,080
Carbon avoided (mt)	628	25,199	28,103

[1] Based on state-wide energy costs for Phoenix: $115.3 per MWH and $16.8 per MBTU; [2] One million British Thermal Units; [3] Megawatt-hour.

3.7. Structural and Functional Values of Phoenix's Park System

The structural value of Phoenix's park system was $692 million with a carbon storage value of $4.53 million. The annual functional value of Phoenix's park system trees was over $1.5 million/year (i.e., carbon sequestration providing $285 thousand/year; air pollutants removal of $1.16 million/year; and energy saving costs combined with carbon emission reduction resulting in $118 thousand/year) (Table 10).

Table 10. City total for trees' structural and functional value by Phoenix's park system.

Typology of Green Infrastructure	Number of Trees (SE)	Carbon Storage (t) (SE)	Carbon Storage Value (US$) (SE)	Carbon Sequestration (t/year) (SE)	Carbon Removal Value (US$) (SE)	Structural Value (US$) (SE)
Urban park system	516,534 (197,947)	57,755.8 (31,396.3)	4,504,952.4 (2,448,911.4)	3633.3 (1288.4)	283,397.4 (100,495.2)	691,812,407 (270,788,843)

SE = Standard error of total.

4. Discussion

Modeling the ecosystem services of Phoenix's urban park system provides not only a picture of the current extent and condition of the system, but also provides a baseline for decision making about strategic interventions into that system that will help the city reach its Tree and Shade Master Plan goals [38]. The modeling approach is a critical first step for decision-makers to understand what ecosystem services urban park systems are currently providing in order to make decisions about how they want to change those services including the efficacy of current strategies (i.e., what is working and not working as intended). As cities implement community greening initiatives, such as Phoenix's Tree and Shade Master Plan, they can replicate the modeling approach to track changes in ecosystem services over time to ensure those initiatives enhance desirable ecosystem services, reduce disservices, and ultimately move the city toward their sustainability goals.

The total number of trees, species and percentage of tree canopy cover are very important elements in assessing the ecosystem services of urban park systems. Ecosystem services are directly proportional to the amount of healthy leaf surface area of the plant [39]. The healthy leaf surface area on individual trees growing on green infrastructure can provide more ecosystem services to citizens. Large trees provide substantially more ecosystem services, such as improving air quality and public health, cooling the air, reducing demand for air conditioning, and supporting climate change adaptation than smaller trees [40]. Trees provide critical climate-regulating ecosystem services by (1) shading surfaces and (2) through evaporative cooling. Tree's leaves reflect from 5% to as much as 30% of the incoming solar shortwave radiation [41]. This reduces the amount of radiation that can be absorbed into the pavement, buildings, and other key contributors to urban-induced heating. Short and long wave radiation play a key role in creating thermally uncomfortable conditions. Trees shield pedestrians from this radiation and provide critical microclimatic regulation of individual comfort. Although there are some large trees within Phoenix's parks (Figure 3), the much larger number of smaller trees may collectively play an important role in providing these ecosystem benefits. The trees growing within Phoenix parks with diameters less than 15.2 cm constitute 53.3% of the tree population (Figure 3), which may suggest that these are relatively young trees and thus likely to be helpful in sustaining the urban ecosystem in Phoenix for years to come. While they are small today, they have the potential to increase in size considerably over time. Alternatively, the dominance of smaller trees may be due to the frequent reliance on desert-adapted tree species including acacia, paloverde, and mesquite. Drought-tolerant native trees such as the paloverde are essentially large shrubs in their native ecosystems [42]. They are adapted to low water use and thus are seen by city officials as more sustainable and water responsible in Phoenix's hot arid climate. Yet, arid cities face key sustainability trade-offs between water conservation or more ecosystem services from larger, more water consumptive trees.

The urban forest cover of 7.2% in Phoenix's park system reduces the heating and runoff impact of impervious surfaces, such as pavement, buildings and, to a lesser degree, maintained turf grass. The 7.2% tree cover in Phoenix's parks compares to Nowak and Crane's [43] finding that Arizona cities had 11.4% urban tree cover in all urban areas. Although small, this tree cover blocks sunlight from impervious surfaces. Impervious surfaces alter the heat energy balance of the land by changing the albedo, sealing soils, and reducing moisture. Impervious surfaces reduce water infiltration and increase runoff, affecting water quality. Trees and vegetation land cover types reduce stormwater impacts by intercepting rainfall, slowing water movement, and increasing infiltration into the ground. Urban park systems hold potential as stormwater infrastructures. Many cities are developing green infrastructure as one important stormwater management strategy. Urban park system vegetative structure can cost-effectively reduce the gray stormwater infrastructure, such as retention tanks. The two most dominant land cover types in Phoenix's park system are bare soil (43.2%) and turf grass (26.6%) (Table 5). These two dominant land cover types are permeable, which means that the park system can be strategically used to control urban stormwater. Phoenix has roughly 64% of its park system available as plantable area and therefore has considerable potential to reduce surface runoff through tree planting initiatives that strategically target the park system. The urban park system is thus an essential component of a city's green infrastructure strategy and should complement other public and private green infrastructure initiatives, such as green streets, landscape ordinances, and school greening efforts as an important stormwater and climate management strategy. An urban park system is a core component of any urban green infrastructure system that can significantly improve the health and sustainability of the local urban ecosystem, providing enduring value for neighborhoods.

Climate change is an issue of global and local concern. Urban trees can help mitigate climate change and meet climate action plan goals by sequestering atmospheric carbon (from carbon dioxide) in tissue and by altering energy use in buildings, and consequently altering carbon dioxide emissions from fossil-fuel based power facilities [37]. Trees store and sequester carbon dioxide through their growth processes in their tissue. As trees grow, they accumulate carbon as wood; when they die and decay they release much of the stored carbon back to the atmosphere. Thus, carbon storage is an indication of the amount of carbon that can be lost if trees are allowed to die and decompose. This makes understanding baseline values a key part of effective long-term management of healthy forests for more holistic climate action planning. Carbon storage and carbon sequestration values are based on estimated or customized local carbon values. Carbon storage and carbon sequestration values in Phoenix are calculated based on $78 per metric ton [29]. The carbon storage in Phoenix's parks are about 57,800 t, with an associated value of $4.53 million (Table 10). Holistic climate action planning can include reuse of dead or green organic waste from urban park systems. Biomass is a renewable energy source that can either be used directly via combustion to produce heat, or indirectly after conversion to various forms of biofuel. Urban park systems are a valuable ecological resource that can be used to provide biomass energy, thus reducing the use of other non-renewable carbon-based forms of energy.

Trees within park systems affect downwind energy consumption by shading adjacent buildings, providing evaporative cooling, thus reducing building energy consumption in the subtropical desert climate with extremely hot summer months. Phoenix reduced energy consumption for residential buildings by around $106,000 annually. Trees within the park system also reduced the amount of carbon released by fossil-fuel based power plants (a reduction of 159 t), with an associated value of $12,482. Phoenix parks represent structural assets with economic value, just as other grey infrastructure in the city. This value is based on the price of replacing existing trees with other similar types of trees. In addition, they also have functional ecosystem service values (both positive and negative) based on the functions the trees perform. The structural values applied here are based on the valuation procedures laid down by the Council of Tree and Landscape Appraisers, which uses tree species, diameter, condition, and location information [22]. The number and size of healthy trees contribute to the increased structural and functional value of an urban forest. These urban forest estimates suggest that the park system offers key ecosystem benefits for city residents today and the potential future for enhancement of those benefits.

Air quality is a major problem in most cities. It can negatively affect human health, ecosystem health, and visibility. Phoenix's park system is the core component of the City's green infrastructure, removing a significant fraction of the air pollution from the city's environment. We found that Phoenix's park system may remove up to 272 tons of air pollutants (CO, NO_2, O_3, PM_{10}, and SO_2) at a value of $1.16 million/year. Air quality and utility managers can use the data to show the benefits of trees and to advocate for more funding to support tree planting and maintenance of tree species that maximize air quality benefits. Strategic tree planting in parks should consider the service provided by trees by locating them near residences to conserve energy, near parking lots to shade pavement, downwind of power plants to reduce pollution, and downwind of congested freeways to reduce vehicular VOC (volatile organic compound) pollution. When people select tree species in green infrastructures, pollutant-sensitive species should be avoided, and plantings should include the use of evergreen trees to improve tree-health and remove particulate matter year-round.

Finally, Phoenix's park system is an underutilized cooling system for many neighborhoods. In particular, many heat-vulnerable residents within the Phoenix valley depend on the cooling ability of urban forests within the City's park system. Although summer temperature in Phoenix on average exceed 37 °C over 92 times and 43 °C over 11 times annually [8], disparities in urban-heating result in dangerous heat vulnerabilities; some neighborhoods are dangerously hotter than others [6]. This exposure to heat is not only during the day, but more critically overnight when heating from urban materials triggers urban heat islands, which lead to locally elevated surface and air temperatures [44]. This exposure to heat increases residents' susceptibly to heat-related illness and death [45]. Low-income households are much more vulnerable to these health effects because the high cost of electricity prevents them from using air conditioning more consistently [46]. At the same time, heat-susceptible residents often do not have access to cooling resources such as air-conditioned cars and must traverse Phoenix's hot streets to use public transportation [6]. Phoenix's urban climate can be cooled by using the City's park system to nurture urban cool islands. Phoenix can use the ecosystem services of its park system to counteract urban heat islands to provide residents better health outcomes for the most susceptible neighborhoods. Urban park systems can improve the air quality and consequently public health, cool the air by counteracting urban heat islands, reduce the demand for air conditioning, and support climate change adaptation, all of which promote the quality of urban life [40]. The environmental benefits provided by the trees in our urban park systems are seldom recognized, but the results of this study suggest that urban park systems are a valuable ecological resource that provide multiple ecosystem services to support sustainable and healthy communities.

The findings of this research fill some important gaps in our knowledge of ecosystem service valuation of urban forests, thus improving the future design, planning and management of green infrastructure in our cities. The assessment of different types of green infrastructure is useful when considering its value in the urban landscape, while understanding an urban forest's structure, function and value can promote decision-making that improves human health and environmental quality. This research captured the current structure of Phoenix's park system as one key part of the city's larger green infrastructure network and quantified a subset of the ecosystem functions and economic values it provides to local residents. The results of this research support better planning and the management of the city's green infrastructure network, providing evidence of the valuable role trees and urban forests play in the places we live by improving the environment and enhancing both human health and the environmental quality of urban areas. The significance of this research is the new vision it provides of urban park systems as land keenly and singularly devoted to nurture more effective ecosystem services. These urban landscapes are valuable ecological resources, enhancing ecosystem health, and promoting a better quality of life for human and non-human city residents.

Findings provide useful information for urban planners, architects, landscape architects, other design professionals, advocates for healthy urban ecosystems, and others concerned with the design and planning of our urban landscape, encouraging them to treat green infrastructure as a life-sustaining resource that improves living conditions and opens up new recreational opportunities for city residents. The methodology that was applied to assess ecosystem services in this study can also be used to assess the ecosystem services provided by green infrastructure in other urban contexts and improve urban forest policies, planning, and the management of green infrastructures.

5. Conclusions

This project addresses broader concepts of sustainability and health by assessing the ecosystem services generated by an urban park system in Phoenix, Arizona. The purpose of this study is to identify and demonstrate how land dedicated to urban park systems can nurture green infrastructure that provides ecosystem services including air pollution removal values, carbon sequestration and storage, avoided runoff, structural value, and energy saving for city residents in the city's hot arid urban climate. The results of this study can be used to inform urban forest planning and management of green infrastructure in any bioclimatic region. It also promotes decision-making that will improve environmental quality and human health and well-being. This study captured the current urban forest structure of Phoenix's park system and its numerous ecosystem services and benefits to Phoenix's resident. The results of this research will support better planning and the management of the city's green infrastructure and are expected to provide data to support the inclusion of trees in the green infrastructure within existing environmental regulations, as well as providing evidence of the valuable role trees play for improving the environment and enhancing both human health and the environmental quality of urban areas.

Author Contributions: G.K. developed research design, analyzed the data and wrote the manuscript. P.C. provided some core advice and review of data and edit the paper. The authors have read and approved the final manuscript.

Funding: This research was supported by funds provided by the North Carolina A&T State University Research Fund and the APS Endowment for Sustainable Design Research Fund (Arizona State University's Herberger Institute for Design and the Arts).

Conflicts of Interest: The authors declare no conflict of interest.

References

1. Assessment, Millennium Ecosystem. *Synthesis Report*; Island: Washington, DC, USA, 2005.
2. Carpenter, S.R.; Mooney, H.A.; Agard, J.; Capistrano, D.; DeFries, R.S.; Díaz, S.; Dietz, T.; Duraiappah, A.K.; Oteng-Yeboah, A.; Pereira, H.M.; et al. Science for managing ecosystem services: Beyond the Millennium Ecosystem Assessment. *Proc. Natl. Acad. Sci. USA* **2009**, *106*, 1305–1312. [CrossRef] [PubMed]
3. United Nations. *Urban Population, Development and the Environment 2011*; Population Division, United Nations Department of Economic, & Social Affairs Population Division, United Nations Publications: New York, NY, USA, 2011.
4. Declet-Barreto, J.; Brazel, A.J.; Martin, C.A.; Chow, W.T.; Harlan, S.L. Creating the park cool island in an inner-city neighborhood: Heat mitigation strategy for Phoenix, AZ. *Urban Ecosyst.* **2013**, *16*, 617–635. [CrossRef]
5. Stone, B., Jr. *The City and the Coming Climate: Climate Change in the Places We Live*; Cambridge University Press: Cambridge, UK, 2012.

6. Harlan, S.L.; Brazel, A.J.; Prashad, L.; Stefanov, W.L.; Larsen, L. Neighborhood microclimates and vulnerability to heat stress. *Soc. Sci. Med.* **2006**, *63*, 2847–2863. [CrossRef] [PubMed]
7. Brazel, A.; Gober, P.; Lee, S.J.; Grossman-Clarke, S.; Zehnder, J.; Hedquist, B.; Comparri, E. Determinants of changes in the regional urban heat island in metropolitan Phoenix (Arizona, USA) between 1990 and 2004. *Clim. Res.* **2007**, *33*, 171–182. [CrossRef]
8. National Weather Service (NWS). 2016; Extreme Temperature Facts for Phoenix and Yuma. Available online: http://www.wrh.noaa.gov/psr/climate.extremeTemps.php (accessed on 21 November 2016).
9. Childers, D.L.; Cadenasso, M.L.; Grove, J.M.; Marshall, V.; McGrath, B.; Pickett, S.T. An ecology for cities: A transformational nexus of design and ecology to advance climate change resilience and urban sustainability. *Sustainability* **2015**, *7*, 3774–3791. [CrossRef]
10. Devitofrancesco, A.; Ghellere, M.; Meroni, I.; Modica, M.; Paleari, S.; Zoboli, R. *Sustainability Assessment of Urban Areas through a Multicriteria Decision Support System: Central Europe towards Sustainable Building*; Grada Publishing: Prague, Czech Republic, 2016; pp. 499–506.
11. Dobbs, C.; Escobedo, F.J.; Zipperer, W.C. A framework for developing urban forest ecosystem services and goods indicators. *Landsc. Urban Plan.* **2011**, *99*, 196–206. [CrossRef]
12. Ciecko, L.; Tenneson, K.; Dilley, J.; Wolf, K. *Seattle's Forest Ecosystem Values Analysis of the Structure, Function, and Economic Benefits*; USDA Forest Service Pacific Northwest Research Station: Portland, OR, USA, 2012; p. 26.
13. Nowak, D.J.; Crane, D.E.; Stevens, J.C.; Hoehn, R.E.; Walton, J.T.; Bond, J. A ground-based method of assessing urban forest structure and ecosystem services. *Arboricult. Urban For.* **2008**, *34*, 347–358.
14. Nowak, D.; Hoehn, R., III; Crane, D.; Weller, L.; Davila, A. *Assessing Urban Forest Effects and Values, Los Angeles' Urban Forest*; Resour. Bull. NRS-47; US Department of Agriculture, Forest Service, Northern Research Station: Newtown Square, PA, USA, 2011; p. 30.
15. McBride, J.; Jacobs, D. Urban forest development: A case study, Menlo Park, California. *Urban Ecol.* **1976**, *2*, 1–14. [CrossRef]
16. McBride, J.R.; Jacobs, D.F. Presettlement forest structure as a factor in urban forest development. *Urban Ecol.* **1986**, *9*, 245–266. [CrossRef]
17. Miller, P.R.; Winer, A.M. Composition and dominance in Los Angeles Basin urban vegetation. *Urban Ecol.* **1984**, *8*, 29–54. [CrossRef]
18. Nowak, D.J. Urban Forest Development and Structure: Analysis of Oakland, California. Ph.D. Dissertation, University of California, Berkeley, CA, USA, 1991.
19. McPherson, E.G. Structure and sustainability of Sacramento's urban forest. *J. Arboricult.* **1998**, *24*, 174–190.
20. Nowak, D.J.; O'Connor, P.R. *Syracuse Urban Forest Master Plan: Guiding the City's Forest Resource into the 21st Century*; US Department of Agriculture, Forest Service, Northeastern Research Station: Newtown Square, PA, USA, 2001.
21. Nowak, D.J.; Crane, D.E. The Urban Forest Effects (UFORE) Model: Quantifying urban forest structure and functions. In *Integrated Tools for Natural Resource Inventories in the 21st Century*; US Department of Agriculture, Forest Service, North Central Research Station: St. Paul, MN, USA, 2000.
22. Nowak, D.J.; Crane, D.E.; Dwyer, J.F. Compensatory value of urban trees in the United States. *J. Arboricult.* **2002**, *28*, 194–199.
23. Ham, D.L. Analysis of the urbanizing of the South Carolina Interstate 85 corridor. In Proceedings of the 2003 National Urban Forest Conference, San Antonio, TX, USA, 17–20 September 2003; p. 67.
24. Nowak, D.J.; Crane, D.E.; Stevens, J.C.; Hoehn, R.E. *The Urban Forest Effects (UFORE) Model: Field Data Collection Manual. V1b*; US Department of Agriculture, Forest Service, Northeastern Research Station: Newtown Square, PA, USA, 2003.
25. Nowak, D.J.; Walton, J.T.; Stevens, J.C.; Crane, D.E.; Hoehn, R.E. Effect of plot and sample size on timing and precision of urban forest assessments. *Arboricult. Urban For.* **2008**, *34*, 386–390.
26. Baldocchi, D. A multi-layer model for estimating sulfur dioxide deposition to a deciduous oak forest canopy. *Atmos. Environ.* **1988**, *22*, 869–884. [CrossRef]
27. Baldocchi, D.D.; Hicks, B.B.; Camara, P. A canopy stomatal resistance model for gaseous deposition to vegetated surfaces. *Atmos. Environ.* **1987**, *21*, 91–101. [CrossRef]

28. Nowak, D.J. *Atmospheric Carbon Dioxide Reduction by Chicago's Urban Forest. Chicago's Urban Forest Ecosystem: Results of the Chicago Urban Forest Climate Project*; Gen. Tech. Rep. NE-186; US Department of Agriculture, Forest Service, Northeastern Forest Experiment Station: Radnor, PA, USA, 1994; pp. 83–94.
29. Interagency Working Group on Social Cost of Carbon United States Government. Technical Support Document: Social Cost of Carbon for Regulatory Impact Analysis under Executive Order 128666. 2010. Available online: http://www.epa.gov/oms/climate/regulations/scc-tsd.pdf (accessed on 20 April 2016).
30. U.S. Forest Service Tree Guides. Available online: http://www.fs.fed.us/psw/programs/used/uep/5ttree_guides.php (accessed on 21 March 2016).
31. Council of Tree and Landscape Appraisers. *Guide for Plant Appraisal*; International Society of Arboriculture: Urbana, IL, USA, 1992.
32. McPherson, E.G.; Simpson, J.R. *Carbon Dioxide Reduction through Urban Forestry*; Gen. Tech. Rep. PSW-171; Pacific Southwest Research Station: Albany, CA, USA, 1999.
33. Kuers, K. Ranking Species Contribution to Forest Community Composition: Calculation of Importance Value. 2005. Available online: http://static.sewanee.edu/Forestry_Geology/watershed_web/Emanuel/ImportanceValues/ImpVal_SET.html (accessed on 19 June 2016).
34. U.S. Department of Agriculture. National Invasive Species Information Center. 2011. Available online: http://www.invasivespeciesinfo.gov/plants/main.shtml (accessed on 28 July 2016).
35. Arizona Wildlands Invasive Plant Working Group. 2005. Invasive Non-Native Plants That Threaten Wildlands in Arizona. Available online: http://www.swvma.org/wp-content/uploads/Invasive-Non-Native-Plants-that-Threaten-Wildlands-in-Arizona.pdf (accessed on 19 July 2016).
36. Kim, G. Assessing urban forest structure, ecosystem services, and economic benefits on vacant land. *Sustainability* **2016**, *8*, 679. [CrossRef]
37. Abdollahi, K.K.; Ning, Z.H.; Appeaning, A. Gulf Coast Regional Climate Change Council. In *Global Climate Change & the Urban Forest*; Franklin Press: Baton Rouge, LA, USA, 2000.
38. Tree and Shade Master Plan—City of Phoenix. 2010. Available online: https://www.phoenix.gov/parks/parks/urban-forest/tree-and-shade (accessed on 7 June 2018).
39. Wiseman, E.; King, J. *i-Tree Ecosystem Analysis Roanoke*; USDA Forest Service Northern Research Station, Trans; College of Natural Resources and Environment: Blacksburg, VA, USA, 2012; p. 27.
40. Rosenthal, J.K.; Crauderueff, R.; Carter, M. *Urban Heat Island Mitigation Can Improve New York City's Environment*; Sustainable South Bronx: New York, NY, USA, 2008.
41. Geiger, R.; Aron, R.H.; Todhunter, P. *The Climate near the Ground*, 7th ed.; Rowman & Littlefield Publishers, Inc.: Lanham, MD, USA, 2009.
42. USDA Plants Database. 2018. Available online: https://plants.sc.egov.usda.gov/core/profile?symbol=PAMI5 (accessed on 13 June 2018).
43. Nowak, D.J.; Crane, D.E. Carbon storage and sequestration by urban trees in the USA. *Environ. Pollut.* **2002**, *116*, 381–389. [CrossRef]
44. Di Sabatino, S.; Leo, L.S.; Hedquist, B.C.; Carter, W.; Fernando, H.J.S. Results from the Phoenix Urban Heat Island (UHI) experiment: Effects at the local, neighbourhood and urban scales. In Proceedings of the EGU General Assembly Conference Abstracts, Vienna, Austria, 19–24 April 2009; Volume 11, p. 12778.
45. Sheridan, S.C.; Kalkstein, A.J.; Kalkstein, L.S. Trends in heat-related mortality in the United States, 1975–2004. *Nat. Hazards* **2009**, *50*, 145–160. [CrossRef]
46. Chow, W.T.; Chuang, W.C.; Gober, P. Vulnerability to extreme heat in metropolitan Phoenix: Spatial, temporal, and demographic dimensions. *Prof. Geogr.* **2012**, *64*, 286–302. [CrossRef]

© 2018 by the authors. Licensee MDPI, Basel, Switzerland. This article is an open access article distributed under the terms and conditions of the Creative Commons Attribution (CC BY) license (http://creativecommons.org/licenses/by/4.0/).

MDPI
St. Alban-Anlage 66
4052 Basel
Switzerland
Tel. +41 61 683 77 34
Fax +41 61 302 89 18
www.mdpi.com

Forests Editorial Office
E-mail: forests@mdpi.com
www.mdpi.com/journal/forests

www.ingramcontent.com/pod-product-compliance
Lightning Source LLC
LaVergne TN
LVHW071952080526
838202LV00064B/6732